DESIGN FOR ENTROPY

DESIGN FOR ENTROPY

ACTIVE INFERENCE AND TECHNOLOGY

LUCA M. POSSATI

The MIT Press
Cambridge, Massachusetts
London, England

The MIT Press
Massachusetts Institute of Technology
77 Massachusetts Avenue
Cambridge, MA 02139
mitpress.mit.edu

The MIT Press would like to thank the anonymous peer reviewers who provided comments on drafts of this book. The generous work of academic experts is essential for establishing the authority and quality of our publications. We acknowledge with gratitude the contributions of these otherwise uncredited readers.

This book was set in Bembo Book MT Pro by Westchester Publishing Services. Printed and bound in the United States of America.

Library of Congress Cataloging-in-Publication Data

Names: Possati, Luca M. (Luca Maria) author
Title: Design for entropy : active inference and technology / Luca M. Possati.
Description: Cambridge, Massachusetts : The MIT Press, [2026] |
Includes bibliographical references and index.
Identifiers: LCCN 2026001055 (print) | LCCN 2026001056 (ebook) |
ISBN 9780262056267 paperback | ISBN 9780262056335 pdf |
ISBN 9780262056342 epub
Subjects: LCSH: Human-computer interaction—Philosophy | Neurosciences
Classification: LCC QA76.9.H85 P666 2026 (print) | LCC QA76.9.H85 (ebook)
LC record available at https://lccn.loc.gov/2026001055
LC ebook record available at https://lccn.loc.gov/2026001056

10 9 8 7 6 5 4 3 2 1

EU Authorised Representative: Easy Access System Europe, Mustamäe tee 50, 10621 Tallinn, Estonia | Email: gpsr.requests@easproject.com

CONTENTS

INTRODUCTION

This book is neither a traditional work of philosophy nor a conventional academic monograph. It does not attempt to establish a "philosophy of active inference and the free energy principle," nor does it delve into classic philosophical debates such as the nature of intelligence, intentionality, or the mind–body problem. Instead, it pursues a clear and specific goal: to examine how active inference and the free energy principle (FEP)—groundbreaking frameworks for understanding cognition and behavior—can deepen our understanding of human–technology interaction and sociotechnical systems. If you are hoping for sweeping theoretical debates about the philosophical foundations of active inference, you might be disappointed.

Yet it would be misleading to say that philosophical issues will be entirely absent. I am, after all, a philosopher of technology, deeply rooted in postphenomenology, (neuro)psychoanalysis, and actor-network theory. My interest in active inference emerged from a desire to bring greater formal precision and empirical grounding to my research. As a philosopher, I see the relationship with science and experimentation not as a side note but as a central aspect of my work. Furthermore, I believe that active inference and the FEP are capable of profoundly changing the basic concepts of psychology (and psychology of technology more specifically) by giving us a new perspective on many problems.

What you are holding is not a technical manual on active inference. If you are looking for an exhaustive neuroscientific, biological, or mathematical breakdown of the theory, this book will not meet those expectations. It was not written by a neuroscientist, though it naturally intersects with neuroscience. Important books and papers of this kind are already available: see, for example, Cheng et al. (2024), Clark (2013, 2014, 2016), Friston (2005, 2008, 2009, 2010a, 2013), Metzinger and Wiese (2017), Nave (2025), Parr et al. (2022), Smith et al. (2022), and Wiese (2017). Instead, this book delves into active inference and the FEP, with a focus tailored to a specialized audience, aiming to go beyond merely popularizing the theory. The goal is to introduce the foundational concepts of active inference and the FEP to social scientists who want to explore how these ideas can enrich their own fields, especially the study of sociotechnical systems and design.[1] It also aims to spark neuroscientists' interest in studying active inference as it relates to human–technology interactions. There is a growing body of literature on active inference and its engineering applications. This book is explicitly outside it. I believe it is necessary to get active inference and the FEP out of a certain technicality that sometimes amounts to a certain academic "confessional" closure.

Let me be more explicit. While there is a substantial body of literature on the relationship between active inference and the social sciences—see, for instance, Badcock et al. (2022), Bruineberg et al. (2018), Constant et al. (2019a), Constant et al. (2019b), Ramstead et al. (2016), Ramstead et al. (2018), Ramstead et al. (2019a), Ramstead et al. (2019b), and Veissière et al. (2020)—and although these works are invaluable, they do not directly address the technological dimension, particularly the question of machine agency. This critical gap becomes evident in their failure to interrogate the relationship between active inference, the FEP, and technology in a holistic sense. This book aims to address this oversight by placing the technological dimension at the forefront of the discussion. It explores technology not merely as the material production of artifacts (i.e., human-made objects) but as a dynamic framework through which human existence is negotiated and renegotiated. Understanding this intricate relationship between humans and technology, as Clark (2016, 294) argues, is "surely one of the major tasks confronting twenty-first-century cognitive science."

While active inference provides a valuable framework, I do not claim it can encompass every dimension of what is often broadly labeled as "technology"—my intent is neither to defend nor to praise active inference, and this is why the book also examines some of the valid criticisms directed at this theory. I see active inference and the FEP as conceptual tools that effectively help us understand certain aspects of human and nonhuman behavior (Baltieri et al. 2020)—but they are, of course, not a theory of everything (Colombo 2019, 2020; Colombo et al. 2021). To be more explicit, from my perspective, the FEP is, first and foremost, a *method* for describing, modeling, and better understanding self-organizing systems. As such, it carries no inherent ontological or epistemological commitments. That said, it can certainly be used to test, support, and develop ontological and epistemological claims. This is exactly how I employ the FEP in this book—for example, when tackling the question of whether we can attribute agency to machines in chapter 3. The FEP offers a method for posing that question and supporting the ontological claim that machines possess agency, but that ontological claim is not intrinsic to the FEP itself. Emphasizing this point is important because the philosophy and social science literatures often (and inappropriately) assume that the FEP is inherently tied to a form of naïve realism or computationalism.

In an era defined by rapid technological advancement, "an accelerating age of machines where processes of computerization and datafication are key drivers" (Lindgren 2024, 4)—where innovations like artificial intelligence (AI), quantum computing, gene editing, blockchain, 5G connectivity, autonomous vehicles, augmented and virtual reality, and nanotechnology are reshaping the world—understanding the cognitive and psychological mechanisms underlying human–technology interaction has become essential. These technologies are not neutral tools; they fundamentally alter how we perceive, learn, and adapt, influencing trust, agency, and behavior in profound ways. Without a deeper understanding of these dynamics, we risk designing systems that alienate users or exacerbate inequalities, rather than empowering individuals and fostering societal progress. As Ihde (1990, 2023) has argued, technology serves as a dynamic mediator between humans and their world, reshaping every facet of human existence—psychological, emotional, intellectual, and practical. Technologies "actively shape the relation between human beings and their world and provide many varied and enriched ways in which that world can be encountered" (Verbeek 2005, 145). Viewing technology exclusively through a technical or engineering lens reduces it to a mere collection of tools, obscuring its profound influence on shaping human identity. This narrow perspective makes us susceptible to the unintended consequences of the very systems we create.

0.1 WHAT IS ACTIVE INFERENCE?

The starting question is straightforward: How do living organisms survive in a world driven by chaos and disorder? The answer lies in their remarkable ability to anticipate challenges and adapt. Rather than passively reacting to their environment, organisms actively predict outcomes and work to minimize errors and uncertainty in those predictions. When discrepancies arise—when reality does not match their expectations—they refine their internal models or take action to align the environment with their predictions. This continuous process of learning from mistakes and improving predictions enables them to thrive in a constantly changing world. There is no fixed or ultimate model of reality; survival depends on the ability to continuously update and optimize these models. This dynamic interplay of prediction, action, and adaptation is the essence of active inference.

Active inference is a mathematical framework derived from the FEP, which describes how self-organizing systems, such as biological organisms, maintain their integrity and adapt to environmental changes. The FEP does not account for the functions or behaviors of systems. Instead, it elucidates why systems persist and maintain their capacity to act, rather than dissipating into disorder. It conceptualizes these systems as Bayesian inference engines that continuously minimize uncertainty or prediction error by aligning internal probabilistic predictions with sensory inputs. This is achieved through two intertwined processes: perception, where the system updates its internal models to better predict sensory data, and action, where it modifies the environment to make observations more consistent with its predictions. Unlike traditional models of cognition that separate perception and action, active inference treats them as codependent, forming a unified process of adaptive behavior. In short, "active inference is a way of understanding sentient behavior" that "puts the action into perception, whereby perception is treated as perceptual inference or hypothesis testing" (Parr et al. 2022, vii). Therefore, it goes "even further and considers planning as inference—that is, inferring what you would do next to resolve uncertainty about your lived world" (Parr et al. 2022, vii).

A common misconception is that the FEP and active inference assume that systems use inference in order to survive. This is a mistake. Free-energy minimization is not a sufficient condition for life or mind (Constant 2020)—only for agency. From this perspective, agency does not necessarily imply life or mind in the full, ordinary sense. The FEP makes no strong assumptions about what qualifies as life or mind; it simply defines some essential features of agency—yet agency itself is not sufficient for life or mind. Thus, attacking the FEP on the grounds that it fails to account for biological life or to distinguish between living and nonliving entities is a conceptual error. The FEP in itself is just "a mathematical statement that says something fundamental (i.e., from first principles) about what it means for a system to exist, and to be 'the kind of thing that it is.' The FEP provides an interpretation of mechanical theories for systems that look as if they have beliefs" (Ramstead et al. 2023, 2), where "beliefs" means probability distributions.

According to the classical scholarship on FEP and active inference, self-organization in biological systems reflects their unique ability to modulate the entropic forces of their environment—in essence, to modulate the dispersive effects of the second law of thermodynamics (or the fluctuation theorem: see Friston 2013, 2019—however, more recent literature on the FEP nuances this connection with the second law of thermodynamics: see Ramstead et al. 2023). This modulation of dissipation is expressed in the ability to minimize a specific quantity—free energy—which serves as a proxy for surprise or *surprisal*; "systems that do not minimize free energy cannot exist, because the entropy of their sensory states would not be bounded and would increase indefinitely" (Friston 2013, 2). Here, "free energy" refers to an information-theoretic measure that is an upper bound on surprisal (statistical unexpectedness) associated with sensory states, given a generative model, that is, a causal model of the organism's surroundings. Living organisms must therefore "minimize the long-term average of surprise to ensure that their sensory entropy remains low" (Friston 2010a, 127)—and they use active inference to achieve this.

Surprise or *surprisal* is just negative log-probability. That's all; there is nothing more mysterious about it. To say that systems minimize surprise is just to say that systems of a certain type behave in type-specific ways (i.e., they move toward more probable configurations, where this is measured by free energy, given a generative model) (for the notion of surprise, see Modirshanechi et al. 2022). Thus, living systems do not "aim to" minimize surprise: given a generative model of their behavior, *we* find observations of their behavior unsurprising.

The FEP experienced a significant conceptual expansion over time. It has been characterized as "a mathematical formulation of how adaptive systems—biological agents, such as animals or brains—resist a natural tendency toward disorder" (Friston 2010a, 127), and as applying to all forms of life, "from single-celled organisms to social networks" (Friston 2009, 293); but also "the free energy principle asserts that for *entities* to maintain their structural and functional integrity, they must change their relationship with their econiche through action" (Ororbia & Friston 2024, 14, my emphasis).

This universalization of the FEP has also been the subject of some criticism, and for good reason. However, there is one crucial point that must be considered: at its core, the FEP is simply a mathematical framework, that is, "a theoretically sound mathematical characterization of the flow of information in the presence of a boundary within a dynamical system."[2] Considering active inference as an empirical theory is a category mistake—it is better framed as an approach to modeling dynamical systems. In other words, it falls under the rubric of mathematics or mathematical physics; "[it] is a mathematical framework, not an empirically testable theory making specific predictions. It can be used to generate testable hypotheses but is not an empirical theory per se."[3] This also means that the application of the FEP is not limited to physics, biology, or brain research—it can justifiably be extended to the social sciences.

The FEP itself does not specify *how* organisms perceive or act upon their surroundings. Active inference, on the other hand, provides an answer to this question by offering a concrete model that directly links the FEP to both perception and action—it is also referred to as a "corollary of the free energy principle" (Kirchhoff et al. 2018). Active inference, unlike passive responses to sensory stimuli, describes how organisms dynamically engage with their environment to minimize free energy. This approach closely aligns with the principles of 4E cognition—embodied, embedded, enactive, and extended mind theories—emphasizing an active, ongoing interaction with the world. Here, agents generate probabilistic predictions about their surroundings and continuously refine them through a feedback loop of perception and action, actively working to minimize prediction errors. This ongoing adjustment aligns the agent's internal states with external realities, allowing it to interpret, navigate, and adapt to its environment in real time.

It is important, I think, to emphasize that active inference and the FEP cannot be reduced to the simplistic notion of "minimizing uncertainty." Instead, they constitute a highly complex mathematical framework that defines a set of quantities and the dynamic equilibria among them—equilibria that can be interpreted in multiple ways. These equilibria reflect some fundamental mechanisms that underlie the behavior of sentient beings. In chapter 2 of this book, I will examine a small yet significant portion of this framework. Failing to consider this aspect risks trivializing these concepts.

0.2 THE THIRD CYBERNETICS

Why explore the connection between active inference and technology? What makes this relationship so important? First, one of the key ideas of this research is that an active inference agent is, by nature, *a technologically oriented agent*. This idea is closely related to FEP. While technology has the potential to enhance our capacity to manage uncertainty, it also tends to introduce novel forms of uncertainty. The central challenge, therefore, lies in co-developing technological systems alongside our cognitive and social capacities to cope with the increased surprisal they entail. A recursive feedback loop between technological advancement and the evolving capacity to bound surprisal may, over time, give rise to a Baldwin effect—culminating, across generations, in the emergence of an increasingly complex technological culture. From this perspective, progress might be conceptualized as a cumulative increase in a species' capacity to bound surprisal over evolutionary timescales. More concretely, the quantity and complexity of information gathered and exchanged among individuals across generations may serve as a useful proxy for tracking the dynamics of cultural evolution.

In other words, *the active inference agent is a designer* because it acts on the world and transforms it—developing (consciously or unconsciously) ideas and creating something new, bringing things into existence by giving them form and structure. I do not mean this metaphorically: the circular causality embedded by the Markov blanket (i.e., the fundamental statistical boundary that structures any active inference agent; more on this later) entails the creative transformation of the environment in accordance with the agent's preferred internal states, and vice versa. This creative transformation is not limited to artifacts in the usual sense but includes organizations, institutions, ideologies, symbols, works of art, and more—technology being just one aspect of this process.

Design is understood here as "devising courses of action that change existing situations into desired situations" (Simon 1981, 129). At the heart of design (and active inference) there is "the ability to imagine that-which-does-not-yet-exist, to make it appear in concrete form as a new, purposeful addition to the real world" (Nelson & Stolterman 2012, 12). Design will always be "about creating something that does not yet exist. It is not about finding something already in existence" (Nelson & Stolterman 2012, 28; see also Parsons 2015.) Even though it is a form of practical rationality, design is grounded in a specific logic. Given a goal, this logic identifies and models a set of requirements, so that some possible systems implementing those requirements qualify as sufficient and feasible solutions to that goal (Floridi 2019). It is a logic of the model of a system, not of the system itself: design is concerned with the project, that is, the informational model of the system to be built, i.e., the blueprint that provides information about the system to be created. It is therefore a future-oriented informational logic. In more philosophical terms, while Kant's transcendental logic looks to the conditions of possibility (the past) and Hegel's dialectical logic to the conditions of (in)stability (the present), design logic concerns the future conditions under which a system can be made feasible.

One of the central arguments of this book is that active inference and the FEP provide an elegant, straightforward theoretical framework for understanding the logic of design. The design and use of any artifact can be seen as an inferential process aimed at the joint minimization of free energy.

Second, active inference and the FEP hold transformative implications across a broad range of fields, from neuroscience and psychology to AI, robotics, and beyond. They offer a physics-based framework for understanding cognition, shedding light on intricate processes like perception, learning, decision-making, and the dynamics of some mental health disorders. This approach not only bridges gaps in understanding natural intelligence but also inspires novel designs for artificial systems. In AI and robotics, active inference informs the creation of adaptive agents capable of autonomous, goal-directed behavior in complex and uncertain environments. By embedding action-perception cycles, these agents dynamically balance exploration and exploitation, enabling them to learn from and adapt to their surroundings. This paradigm can be crucial for developing technologies that seamlessly integrate with human needs and environments. Moreover, the applicability of active inference extends beyond individual agents to sociotechnical systems. This has already sparked groundbreaking innovations in fields such as smart city governance, where predictive models optimize resource allocation and environmental adaptability, as well as human–robot interaction, improving collaboration and communication.[4]

Third, in more specific terms: Why should designers and engineers care about active inference and the FEP? Because they give them a normative design framework that unifies perception, decision-making, and control. In this framework, perceptual inference updates the agent's beliefs to reduce variational free energy, while action selection and planning aim to reduce expected free energy over future horizons—picking policies that both drive preferred outcomes and actively shrink the uncertainties that matter for control. In other terms, when you formalize an artifact as an active-inference agent, you obtain a within-model, Bayes-optimal account of its behavior—given the specified generative model, preferences over outcomes, and chosen variational approximation—subject to resource constraints. In practice, the agent updates beliefs via variational Bayes and selects policies by minimizing expected free energy, that is, maximizing expected model evidence, using a transparent criterion grounded in the standard variational bound. Preferences are expressed as distributions over observables, which implicitly define a viability set. By minimizing the expected free energy, the agent tends to avoid undesirable states, limit ambiguity, and seek information when useful (epistemic value), helping it remain within that viability set over time. By contrast, reinforcement learning optimizes expected return; viability must be engineered (e.g., constraints, penalties, or risk-sensitive objectives), whereas in active inference viability-preserving, uncertainty-aware behavior follows directly from the core decision rule. In this way, active inference and the FEP offer designers a formal model of sustainability, understood as maintaining a system within its viability set over time.

But it is not just this. Beyond these three primary reasons, there is an additional, deeper motivation that drives me to examine the intrinsic relationship between the FEP and technology. This is connected to a broader cybernetic trajectory within the active inference literature, which emphasizes the substantial continuity between living organisms and machines. Recent foundational works such as Ororbia and Friston (2024), with their concept of "mortal computation," and Friston et al. (2024, 2), that propose the framework for "a cyber-physical ecosystem of natural and synthetic sense-making in which humans are integral participants—what we call 'shared intelligence'" exemplify this approach. These contributions highlight the profound interdependence, that is, "shared intelligence," of biological and synthetic systems within the active inference paradigm. Is this enough to speak of a cybernetics revival, or even a third wave of cybernetics?[5] Are we witnessing a resurgence of cybernetics, revisiting its original research agenda despite its early—and perhaps only apparent—failure (Dupuy 2009)?

In its immense ambition to unify physics, biology, psychology, neuroscience, information theory, and AI, active inference stands as the last great heir of the original cybernetic project—namely, the effort to establish a physicalist science of the mind and thereby resolve the age-old philosophical problem of how mind and matter relate. This means, in other words, that active inference revives two foundational ideas from early cybernetics (Wiener, Rosenblueth, Bigelow, McCulloch, Pitts, von Neumann, Ashby): (1) that thinking is a form of computation—though not merely symbolic computation but modeled on McCulloch's neural networks ("man-made machines are not brains, but brains are a very ill-understood variety of computing machines. Cybernetics has helped to pull down the wall between the great world of physics and the ghetto of the mind" [McCulloch 2016, 163]; this is not about anthropomorphizing machines but about *mechanizing humans*: see Dupuy 2009)—and (2) that physical laws can account for why and how the world seems infused with meaning, purpose, directionality, and intentionality—the idea of a physics of meaning (Petitot 1992) strongly resonates in the strand of active inference literature centered on Bayesian mechanics, "a physics of and by beliefs" (Ramstead et al. 2023).

0.3 SCOPE AND STRUCTURE OF THE BOOK

The book is structured into six chapters. Chapters 3 and 4 serve as a transition point, bridging the more theoretical foundation laid in the first two chapters with the practical applications explored in the final two. As said, this volume is not a design manual or a technical tutorial. It is a theoretical monograph that proposes a unified framework—grounded in predictive processing, active inference, and the FEP—for making sense of technology and design practices. Therefore, it is intended as an introduction and orientation to more substantial empirical research to follow—some books look back and take stock, while others look ahead and lay out roadmaps for the future; this book falls into the latter camp. The cases (smartphones, robotics, cities, holographic sensing) are conceptual rather than prescriptive; they are meant to show how artifacts mediate the synchronization between the generative models of designers and users. The book offers a vocabulary and a conceptual map for scholars in philosophy of technology, science and technology studies (STS), user experience design (UX), and cognitive science; it does not provide experimental protocols or detailed operational metrics, which will be developed in future work. Readers will therefore find arguments, theoretical connections, and research avenues; they will not find "turnkey" design checklists.

Chapters 1 and 2 of the book provide a general introduction to active inference and the FEP. This introduction unfolds on two levels: a shorter, more concise overview (chapter 1) focused on predictive processing (PP) and a more technical, in-depth discussion about active inference (chapter 2) that also examines critical points and novel developments. PP can be seen as an application of active inference to neuroscience—what Parr et al. (2022) refer to as "the low road" to active inference. Readers primarily interested in practical applications in design and the psychology of technology may choose to read only chapter 1 before skipping directly to chapters 5 and 6—without missing any key concepts. However, for those who also wish to explore the theoretical foundations of active inference, chapters 2 and 3 are essential. Let me point out again that chapter 2 is *not* a mathematical-engineering tutorial on the FEP and active inference, nor a lightweight, popular overview. Rather, it tries to strike a careful balance—detailed enough to reveal the full depth and elegance of the framework, yet accessible enough. I chose to introduce a certain level of formalism to clearly illustrate the physical and mathematical foundations of active inference, as doing so underscores the relevance of this approach.

Chapters 3 and 4 link the book's theoretical foundation (chapters 1 and 2) with its hands-on focus (chapters 5 and 6). Chapter 3 unfolds in two parts. In the first part, I survey current social-science applications of active inference and the FEP, offering a targeted critique: these works have overlooked the pivotal influence of technology and, in particular, the problem of artifact agency. Assigning agency to objects and artifacts is a crucial issue in current technology-studies debates—and how we answer this question has wide-ranging implications not only for the philosophy and ethics of technology, but especially for design. I claim that the FEP provides an elegant framework for justifying the attribution of agency to artifacts—this is precisely the added value of incorporating the FEP into technology studies.

My argument is as follows. Within the FEP framework, agency is fundamentally self-evidencing—it naturally arises in any system that possesses a Markov blanket and a generative model. This means that the existence and type of Markov blanket are the key structural conditions under which a system admits an interpretation as a (minimal) self-evidencing agent—a nontrivial fact, since one cannot arbitrarily designate any system as having a Markov blanket (Beck & Ramstead 2025; Kirchhoff et al. 2018; Sakthivadivel 2022a). Furthermore, agency should be viewed along a continuum: its intricacy and temporal extent vary according to the architecture and complexity of the Markov blankets and generative models of the agent. Now, this perspective offers a versatile, physics-based conception of agency that applies even to artifacts, so long as they truly exhibit a Markov blanket and a generative model. This thesis is further reinforced by the capacity to reinterpret and refine central concepts in the philosophy of technology, design, and cognitive science—such as technical mediation and symmetrical ontology—using the FEP, as demonstrated in the second part of the chapter.

Chapter 4 further extends the argument of chapter 3 by laying the groundwork for an FEP-based design methodology. Its central thesis is that design consists in engineering Markov blankets. This is far from obvious. A Markov blanket is not an arbitrary boundary; it is the minimal set of variables that renders internal and external states conditionally independent, thereby defining their statistical separation. In active inference, this blanket decomposes into sensory and active states, which realize the interfaces between the internal and external states. You cannot merely label a box or widget with "Markov blanket"—you have to demonstrate that it actually structures inference in the way the formalism requires. Building on Beck and Ramstead's (2025) unsupervised Markov blanket detection algorithm, the chapter shows how designers invert that model: by embedding informative priors over prospective boundary variables, executing targeted, free-energy–minimizing interventions (e.g., rewiring connections, adding sensors, rearranging components), and focusing computational attention on candidate separators, they can steer a complex system to self-organize into a new autonomous agent. Inverting Beck and Ramstead's (2025) algorithm to engineer new blankets embeds design practice directly in the physics of self-organization, rather than treating "design" as a metaphor. This design-as-inference protocol unfolds through interwoven levers—precision shaping (modulating the weighting of sensory inputs via graphical, material, or haptic affordances), curiosity sculpting (introducing controlled uncertainty to drive epistemic exploration), and prediction embedding (externalizing expectations in artifacts—which accelerate convergence toward a clear statistical boundary). These levers are design knobs that modulate active inference and self-organization.

The second part of chapter 4 analyzes the relationship between artifacts and end users. Crucially, artifacts serve as mediators of deferred synchronization in a triadic loop—designer, artifact, user—by embodying the designer's generative model and enabling predictive alignment across time and space. Finally, the chapter focuses on innovation and groups. It introduces the notion of "area of technological innovation"—built on the "zone of bounded surprisal" (Manrique et al. 2024)—as the leadership-driven domain that channels initially "out-of-bounds" inventions into collective practice until saturation or open-ended evolution, raising the group's cognitive threshold and fueling successive innovation cycles.

Building on the theoretical foundations laid in earlier chapters, chapter 5 explores four case studies—smartphones, robotics, smart cities, and holography with dense wireless networks. The chapter offers a conceptual analysis of the dynamic interactions between designers, artifacts, and users, identifying the Markov blankets and generative models that shape their relationships in the cases. The appendix introduces the Designer-Artifact-User (DAU) software package. The DAU is a prototype design-tool concept that designers can use to create, refine, and showcase projects across a range of fields. What sets DAU apart is its foundation in active inference. Using the DAU, the designer can represent their artifact as an active-inference agent and then simulate its interactions with the end user—also modeled as an active-inference agent, complete with its own generative model, priors, Markov blanket, and other characteristics. This setup lets the designer view the design process as a dynamic, mutual adaptation between agents, providing deeper insight into how their choices shape the outcome. Unlike traditional user experience design (UX) models that rely on post-hoc data and user testing, DAU simulates how artifacts embed predictive expectations about their use and users, allowing designers to anticipate mismatches between intended design and real-world experience.

Chapter 6 delves into the psychology of technology, investigating how active inference sheds light on fundamental psychological concepts while addressing technology-related pathologies. Consequently, I analyze certain psychological issues and interpret them through the lens of active inference—while, of course, taking into account existing studies that have already explored this approach. The chapter unfolds in three interconnected parts. The first part lays the theoretical foundation, redefining core psychological concepts through the lens of active inference and the FEP. Drawing on the seminal work of Feldman Barrett (2017) and, in particular, Solms (2021), this section shows how these frameworks provide a deeper understanding of emotions in technology use. The second part applies this theoretical lens to pressing contemporary challenges, analyzing the psychological pathologies linked to technology use, with a focus on conditions such as depression and addiction. The third part focuses on the concept of mentalization, as developed by Bateman and Fonagy (2016). I argue that mentalizing—the ability to attribute mental states such as intentions, beliefs, and desires to others—plays a fundamental role in human–technology relationship. The difference between an artifact and a natural object lies in the process of mutual mentalization between users and designers, which defines the function—not the agency—of an artifact. The designer embeds an implicit understanding of the user's mind into the artifact, anticipating how they will interpret and interact with it. Conversely, the user recognizes the artifact as a product of human intentionality, attributing a mental state to the designer through their engagement with the object. In other words, an artifact's function is not a mysterious metaphysical property; it is determined by how well its users and designers mentalize one another—that is, by the extent to which those who employ the technology for their own ends and those who created it for theirs understand and anticipate each other's mental states.

Now, active inference gives us a highly persuasive theoretical framework for explaining mentalization, redefining it as the synchronization of generative models (Friston and Frith 2015a, 2015b). The discussion will explore three key points: (1) Mentalization can be framed through active inference, where agents minimize uncertainty about each other's intentions; (2) synchronization of generative models is a necessary but insufficient condition for mentalization, which is a more complex process; and (3) mentalization itself is a prerequisite for artifacts to exist as designed entities rather than mere physical objects. The advantage of framing mentalization within the framework of active inference lies in its explanatory power: it accounts for how three agents—artifact, designer, and user—converge on shared generative models without requiring predefined external rules. Instead of relying on explicit communication, designer and user implicitly coordinate through the artifact itself, making technology a deeply cognitive phenomenon rooted in our capacity to understand and predict each other's minds.

1

THE BRAIN AS A PREDICTIVE MACHINE

Perceiving the world around us may feel effortless, but this ease is an illusion crafted by our brains. In truth, perception is an intricate process that depends on a specialized form of inference and is fundamentally tied to action. Moreover, every perception is uniquely shaped by an individual's experiences—beliefs, knowledge, memories, and even fantasies (Clark 2014, 2023; Frith 2007). At the heart of this process lies the brain's most essential tool: prediction. According to PP, the brain is a highly complex predictive machine, and what we see, feel, and hear is "nothing more than the brain's best guess at the causes of its sensory inputs," and that is why the content of our experience is a kind of "waking dream, a controlled hallucination, and that is both more and less than whatever the real world really is" (Seth 2021, 76). This simple idea seems completely counterintuitive. Common sense tells us something different: there is a mind-independent reality, an apple on the table, sensory data comes to my brain through my senses, and finally my brain processes this data by creating an internal "representation" of the apple—as a little internal "spectacle." This is the bottom-up perspective: from the senses to the brain—not the other way around. The brain has no creative role in this process. It simply has to process the information coming from the surrounding reality. This is the classical view even in computer science and artificial intelligence (Marr 1982 remains the main reference in this respect; for a critical history of neuroscience, see Chirimuuta 2024). PP reverses this perspective: our experience is the result of the encounter between bottom-up and top-down activities in a hierarchical space. In this encounter, top-down activities are the predictions that the brain produces and play an active role in shaping the content of our experience. This means that the moment I look at the apple on the table, my brain is already anticipating what will come next; it is making predictions about the apple and its surroundings in general. These predictions are gradually tested, that is, compared with the perceptual material and then selected.

The brain focuses on the error: this is really informative. The error is the discrepancy between the predictions and the sensory data. The error enables the brain to determine which prediction is most accurate. The crucial idea of PP (originally formulated by Hermann von Helmholtz and Richard Gregory) is that perception is the result of an unconscious inferential process, in the sense that the content of perception is not given but is constructed and constantly updated by a particular kind of probabilistic inference. Therefore, "perception happens through a continual process of prediction error minimization" (Seth 2021, 83). It is a controlled hallucination[1] (i.e., tested by comparison with sensory data), just as hallucination is an uncontrolled perception.

Predictive processing can be translated into neurological terms, into "the sparse connectivity structure of biological computation" (Parr et al. 2022, 85; see also Parr & Friston 2018). Neural activity encodes predictions, that is, probability distributions over states of the world that cause the sensory inputs. From this perspective, PP provides a theoretical framework that allows us to develop testable hypotheses using empirical data. This means that these concepts can be translated into the language of synaptic communication, that is, a dynamic message passing between lower and higher levels of the cortical hierarchy (see Parr et al. 2022, chapter 5; for the versatility of PP models to reproduce and analyze neurophysiological data: see Smith et al. 2022). In this hierarchical framework, higher-level predictions help suppress or resolve discrepancies at lower levels, a phenomenon called repetition suppression (Murray et al. 2002; Summerfield et al. 2008). The balance between sensory input and predictions is thought to be regulated by several neural mechanisms, encoding the reliability or precision of these signals—a process known as precision weighting (Friston 2009, 2010a, 2010b, 2017). It is important to note that in active inference, precision is a technical term that refers to the confidence (inverse variance) assigned to prediction errors or beliefs. Agents can modulate precision to change how strongly different sources of information influence inference and action. This modulation is typically associated with attention, but can also be linked to emotional states, which shape the degree of confidence in particular predictions.

PP is now a prominent theoretical framework in the philosophy of mind and cognitive science (Mendonça et al. 2020; Metzinger & Wiese 2017). This is mainly because of its multiple connections to other disciplines (computational neuroscience, biology, machine learning, AI, psychology, psychiatry) and the resources it offers to unify different issues (perception, cognition, action, psychopathology, learning) and thus enable interaction between disciplines. PP stands as "a promise for a unique theoretical background that will unify the various sciences that study the brain, the mind, and its psychology into one systematic framework" (Mendonça et al. 2020, 1). However, it is important to note that there are multiple perspectives on PP, such as Hohwy's (2013) more neurocentric and internalist approach and Clark's (2013, 2016, 2017, 2023; Clark et al. 2012) embodied, action-oriented framework.

In this chapter, I aim to provide only a general overview of PP, primarily following what Clark (2013) refers to as "radical predictive processing." It is also important to emphasize that PP should not be conflated with active inference. PP refers to a family of theories that are only partly related to the FEP.

1.1 THE LOGIC OF PREDICTION

The foundational idea of PP is that the brain maintains an internal model of reality—a probabilistic simulation that is continuously updated based on mismatches between predictions and incoming sensory data. The purpose of this simulation is to anticipate reality, specifically sensory inputs. This internal model is referred to as a *generative model* because it actively generates data, producing simulations and predictions. The generative model has a bidirectional hierarchical structure, meaning that top-down predictions self-generate, encounter, and shape bottom-up signals (figure 1.1). As said, in this dance, what truly matters is the mismatch—what we call the error or unexpected variation. This is what we actually refer to as experience. In other words, the brain is Bayesian (more on this later).

To explain this idea, Clark (2013) draws on predictive coding, a data compression technique: "consider a basic task such as image transmission: In most images, the value of one pixel regularly predicts the value of its nearest neighbors, with differences marking important features such as the boundaries between objects. That means that the code for a rich image can be compressed (for a properly informed receiver) by encoding only the unexpected variation" (Clark 2013, 182). The entire activity of our brain is nothing more than a continuous cycle of predictions and corrections of those predictions through the detection of errors. Therefore, "for my brain, perception is a loop" (Frith 2007, 126), meaning that perception begins within the brain

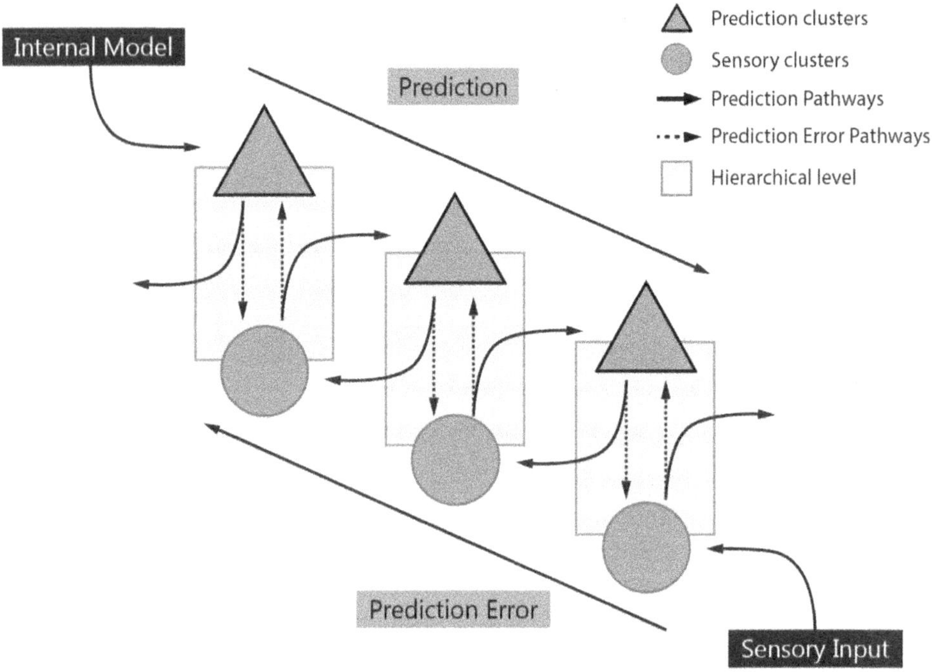

Figure 1.1
Simplified sensory prediction model: higher levels send predictions to lower levels, which compute prediction errors and pass them upward across hierarchical processing stages. *Source:* C. Fong, W. Law, T. Uka, and S. Koike, 2020, "Auditory Mismatch Negativity under Predictive Coding Framework and Its Role in Psychotic Disorders," *Frontiers in Psychiatry* 11, p. 557932, https://doi.org.10.3389/fpsyt.2020.557932.

itself, originating from internally generated and encoded predictions. It then unfolds through a comparison between these simulations and incoming sensory data. "These errors teach my brain to perceive. The existence of the errors tells my brain that its model of the world is not good enough" (Frith 2007, 126–127). This is both the most economical and the most effective way to process sensory inputs. In the PP model, the information that needs to be communicated "is just the prediction error: the divergence from the expected signal" (Clark 2013, 183). The goal of this process is not to achieve a perfect match between the brain's internal states and the external world. Rather, the objective is to infer the causes of incoming sensory input by evaluating its variability in order to enhance survival. The brain—essentially a black box, isolated from the real world and enclosed within the skull—can do nothing but learn to guess more and more accurately. Empirical evidence supports this thesis (see, for instance, Bastos et al. 2012; Friston 2010b; Isomura et al. 2023; Parr & Friston 2018; Zmarz & Keller 2016).

As we will see in the next chapter, active inference and the FEP provide a mathematical foundation for PP. Minimizing free energy is essentially equivalent to minimizing prediction errors.

We cannot treat all prediction errors equally. If we did, there would be no learning, no selection of content based on context—just a constant, undifferentiated flow of discrepancies between probabilistic predictions and data. Therefore, we need to introduce another element: the evaluation of prediction accuracy, also known as the *precision weighting mechanism* (Clark 2023, 224–228). This adds a meta-level to our analysis. The brain not only predicts incoming sensory data but also evaluates the reliability of its predictions and the resulting prediction errors. It "has to select the channels it listens to—by adjusting the volume or *gain* of prediction errors that compete to update expectations in higher levels"; computationally, "this gain corresponds to the precision or confidence associated with ascending prediction errors" (Friston & Frith 2015b, 392). In more technical terms, "precision quantifies the confidence associated with a given probabilistic distribution" (Parr & Friston 2018, 12). This approach makes predictions and errors more sensitive to context. For example, if I am in a club where the music is very loud, my brain gives little relevance to most predictions made in that environment, instead choosing to focus on only a few of them—those that are more reliable because they are based on prior knowledge, such as familiarity with the location. For this reason, improving or updating predictions must also include evaluating the accuracy of both the predictions and the errors.

As mentioned, according to PP scholars, the precision weighting mechanism is closely linked to the function of attention, that is, optimization of expected precisions in hierarchical prediction error minimization (Clark 2013; Feldman & Friston 2010; Friston 2009; Sandved-Smith et al. 2021). Attention "might not be the 'selection' of sensory channels but an emergent property of 'prediction'; where high-precision prediction-errors enjoy greater gain" (Friston 2009, 299). Basically, the brain gives more weight to sensory information when it is considered very accurate and reliable, and less weight when it is uncertain. Precision estimates serve as weightings for prediction error terms, determining how strongly a prediction error influences processing at other levels of the hierarchy (i.e., how informative it is).

(From a neurobiological perspective, various brain mechanisms are linked to the precision of beliefs; see Parr & Friston 2018.)

When the estimated precision of a prediction error increases, the corresponding sensory data receives deeper processing. Precision estimates can be updated through both bottom-up and top-down mechanisms. Bottom-up, precision can be calculated from sensory data, for instance, as the inverse of sample variance derived from observed inputs. Top-down, precision estimates can be modulated based on contextual expectations, such as anticipating increases or decreases in precision, or they can act as goal representations that guide mental actions (Metzinger & Wiese 2017). Imagine walking around a well-lit room. Your brain predicts where objects are based on vision because visual information is accurate and reliable. On the other hand, if you walk in the same room in the dark, your brain gives more weight to tactile information (such as touching furniture) because visual information is no longer reliable. In this way, the brain uses the precision weighting process to adapt to different situations and ensure that you move safely and effectively. Therefore, the precision weighting process helps the brain effectively navigate different situations and environments, optimizing responses to changing conditions. Hohwy (2013, 193) highlights that "engaging in predictive inference without taking account of precision would be as precarious as engaging in statistical comparison of means without taking the variance about the means into consideration." Having samples with low variance enhances statistical reliability, leading to more accurate and robust models. However, different states of the world come with varying levels of noise and uncertainty, which is why variance can be problematic. That is why—according to Hohwy (2013, 2020)—the predictive brain evaluates the accuracy of both the data and the predictions by making second-degree predictions—these are predictions not only about the accuracy of initial predictions but also about the accuracy of prediction errors connected to that prediction. This process involves multiple layers of first- and second-degree predictions.

1.2 THE BRAIN IS NOT AN ORGAN PASSIVELY WAITING FOR STIMULI

Let us take a step back. How does the brain generate predictions? What does it mean for the brain to predict? How does the brain minimize prediction errors? There is a foundational premise of PP that cannot be overlooked, which has been well described by Feldman Barrett (2017). The passage is quite lengthy, but I find it truly fascinating.

> Your brain's 86 billion neurons, which are connected into massive networks, never lie dormant awaiting a jump-start. Your neurons are always stimulating each other, sometimes millions at a time. Given enough oxygen and nutrients, these huge cascades of stimulation continue from birth until death. This activity is nothing like a reaction triggered by the outside world. It's more like breathing, a process that requires no external catalyst. (Feldman Barrett 2017, 58)

Our neurons are constantly active, communicating with each other and simulating data. This form of interaction and collaboration is referred to by Feldman Barrett (2017, 58) as the "intrinsic brain activity," where neural networks operate like sports teams. This allows us to highlight a crucial point. The sensory data we typically think of as coming exclusively from the external world is, in fact, much more complex. The majority of the content our brain processes is generated by the brain itself—neurons communicating with other neurons, simulating certain situations, and testing these simulations against reality. On average, throughout the cortex, downward (inside to out) pathways outnumber feed-forward pathways. This fact is so fundamental that we often struggle to recognize it: we continuously produce internal images, sounds, and sensations (the most obvious examples being dreams, hallucinations, and imagination) that blend with the data actually coming from outside the brain. Most of our experience is a cerebral simulation aimed at predicting what will happen next, anticipating consequences, and calculating the error in comparison to the data. In short, the human mind is a set of hierarchically interconnected predictions that differ only in their object (proprioceptive, exteroceptive, interoceptive) and their temporal depth (present, near future, distant future). Prediction error arises from the comparison between these levels.

If the brain operated solely as a reactive system, it would be far too inefficient to support human survival. The sensory input it processes is immense and relentless (Dimakou et al. 2025). For example, a single retina transmits visual information at a rate comparable to a fully loaded computer network connection every moment you are awake. Now imagine the sheer volume of data generated by all your sensory pathways combined—it is staggering. A reactive brain would quickly become overwhelmed, much like a transportation network grinding to a halt under the strain of excessive traffic. Beyond inefficiency, a purely reactive brain would be metabolically unsustainable, requiring an unfeasible number of neural interconnections to manage such vast input. Evolution, however, optimized the brain for efficiency by wiring it to prioritize prediction over reaction (Feldman Barrett 2017, 60–61).

1.3 ACTION AND PERCEPTION

A key aspect of certain approaches to PP is the role of action. The minimization of prediction errors is achieved through action, highlighting not only the deep integration of brain and body as a unified whole but also the essential role that the body and its (external and internal) movements play in updating predictions (Clark 2023; Metzinger & Wiese 2017, 22). The brain's primary goal is to help us survive in a complex and unpredictable world. To do this, it relies on predictions, prediction errors, and precision estimates, but also on actions with which it can transform sensory data. A PP agent acts freely in the world to transform it, thereby altering perceptual data to align with its predictions (Clark 2023; Nave 2025; Nave et al. 2020). A classic example is the scenario of feeling cold in a room despite expecting warmth. In response, the agent gets up and closes the window, thus confirming its predictions—at that point, the sensory input matches the expected outcome.

However, it is important to emphasize the dual value of prediction error minimization. As Clark writes, actions "are caused by predictions of the flow of the bodily sensations that would occur if that very action was to be performed" (Clark 2023, 70). In other words, when the brain initiates an action, it generates a cascade of predictions—but an error arises. At that point, the error triggers a new prediction concerning the sensory consequences of a possible action—how, through an action, to modify the data to better match the initial predictions. This connection between learning and planning reverses the usual understanding of causality (the representation of an effect triggers the action intended to produce it) and aligns with the ideomotor theory of action, originally developed by William James—actions are initiated by imagining and anticipating their sensory outcomes (Clark 2013, 186; Clark 2014; Clark 2023, 71; Frith 2007; Hohwy 2013; Metzinger & Wiese 2017, 18–20). Thus, prediction error minimization is both the cause and the purpose of action. On one hand, prediction error generates new predictions about bodily movements and how such movements are aimed at changing the sensory flow. On the other hand, these predictions—the idea of the action—activate the motor commands that produce the movements (see also Friston 2011). A classic example that illustrates this point is that you cannot tickle yourself. The prediction nullifies the surprise, eliminating the effect, because you know exactly what is happening (Clark 2016, 112–115; Clark 2023, 78–80).

But there are two problems. The first is: How does my brain know which movement to perform in order to reduce the error of the initial predictions? The solution is fairly simple: the brain learns to do this gradually by acquiring a learned predictive model. "We can think of that model as the distilled wisdom from prior experience" (Clark 2023, 74). By using the learned predictive model, the brain can immediately eliminate the error and initiate a cascade of predictions, making them come true. The hidden task of all that training is "to enable our brains to predict . . . we learn how things will look and feel if we are getting the action right—we learn its sensory consequences" (Clark 2023, 84). For example, if I hear the sound of a siren on the street, I predict that I will see an ambulance passing by when I look out my window. To make this prediction come true and eliminate the error resulting from the fact that I cannot see the ambulance while sitting in my chair, I simply get up and go to the window. The action eliminates the error and confirms the initial prediction.

The second problem is much more complex and directly concerns the causality of action. If the brain's goal is to reduce prediction error based on incoming sensory data (whether from inside or outside the body), then how is movement possible? If the brain predicts future actions, but this prediction conflicts with current sensory data showing no movement is occurring, the logical result is the correction of the error, which in turn nullifies the initial prediction. For example, if I want to grab my cup, my brain generates a series of predictions about the movement required to grasp that cup. However, the sensory data at that moment contradicts those predictions because I am not yet moving to grab the cup. Therefore, the brain should automatically discard those predictions as incorrect and update the generative model—the fact is, my hand is not moving. This creates a paradox: if the brain's mechanism for minimizing prediction error works as described, it should continually cancel out any predictions of future movement that are not immediately reflected in current sensory data, making intentional movement impossible.

The solution lies in the precision-weighting mechanism. The brain assigns greater importance (attention) to predictions concerning future bodily movement than to those concerning the current state of the body. Consequently, as Clark explains, to move my arm, "I must give high weighting to the desired future state (arm moving) rather than the undesired present state (arm not moving). This means actively disattending to the present state of the arm, hence dampening down sensory information" (Clark 2023, 77). This mechanism, called "sensory attenuation," helps reconcile the paradox of initiating and sustaining intentional actions while maintaining a coherent and adaptive response to the environment (Friston & Frith 2015b; Pezzulo 2013; Seth 2015). The minimization of prediction errors is the primary driver of movement and behavior in general. This also creates a distinction between two levels of predictions based on temporal depth: predictions about the future and predictions about the present. Predictions about the future are broader and more abstract, concerning sequences of actions, and are considered more important in the sense that they—and the errors associated with them—receive greater precision. Future-oriented predictions, therefore, filter present-oriented predictions, meaning that present predictions that contradict future predictions are given less attention. Additionally, future predictions have a counterfactual structure: the brain evaluates what would happen (what sensory stimuli it would receive) if those predictions were to come true. This counterfactual evaluation allows the brain to prioritize the future state over the present, enabling movement to occur despite initial sensory contradictions.

One important point must be stressed. As said earlier, the precision-weighting mechanism can be understood as a meta-predictive layer. It is plausible to distinguish between two intertwined types of precision in the PP perspective: (1) goal-related precision, which prioritizes predictions and errors based on their relevance to achieving the intended outcome, and (2) variability-related precision, which reflects the instability and uncertainty of a given prediction or sensory data errors affecting highly significant predictions—those considered very reliable—which can have system-wide effects, forcing substantial changes throughout the brain. A "desert landscape" situation is also possible (Clark 2013; Friston 2010a). This term refers to a state where the brain's generative model lacks the precision needed to minimize prediction errors effectively. In this state, sensory input and predictions fail to converge, leading to overwhelming uncertainty, disorientation, anxiety, and an inability to assign meaning to incoming information.

This allows us to clarify another relevant point. PP asserts that there is an active codependence, a causal circularity, between action and perception—"perception and action constantly engage *in a kind of coupled unfolding*" (Clark 2016, 121–123, my emphasis; see also Clark 2023, 75; Nave 2025). We use our bodies to perceive and understand the world. Predictions, therefore, primarily concern the body and its movements. My perception "is a prediction of what ought to be out there in the world. And this prediction is constantly tested by action" (Frith 2007, 132). When an error arises, the brain has two options. The first is to change the predictions by generating new ones or updating already existing generative models based on new information, that is, the errors. The second option is to change the flow of data by acting directly on the world—gathering new sensory data that better aligns with the current predictions, thereby minimizing the error. For example, if I expect the room to be quiet but it turns out to be noisy, I can either reduce the noise by closing the door or adapt to the situation by putting on noise-canceling headphones. In both cases, the action alters the flow of data to match the initial predictions and minimize the error, reducing the uncertainty related to those predictions. However, this may not always be possible. When it is not, as mentioned, the predictions themselves must change—altering how we process information and construct our predictions. This idea is fundamentally different from the classic sense-think-act model, in which perception and action are two distinct processes—first, perception gathers data, and then the brain plans the actions to be taken. In that model, action is merely the execution of a plan. PP completely overturns this model.

Consider a baseball player who needs to catch a ball hit by the batter. Their actions are an intricate set of predictive-epistemic and pragmatic actions, combined with different types and degrees of precision estimation. Predictions serve to control the execution of actions and solve the problem at hand. Precision estimation, on the other hand, acts as a filter to select actions that improve the process on different timescales. The process is hierarchical: there are predictions that focus on the near future (the goal of the action, the layout of the field, the movements of other players) and predictions with a more limited scope, concerning micro-movements that gradually adjust the hand's position relative to the ball. The movement of the eyes gathers essential information to continuously update predictions about the movements of the arms and legs. This process is supported by cascades of internal predictions regarding the state of the organs, or internal micro-actions aimed at tuning the organs to the overall physical effort (e.g., increasing heart rate or adjusting body temperature.). Cascades of predictions correspond to cascades of errors, which are gradually eliminated through new predictions and thus new actions. The goal is to refine the model. But the model is not an abstract idea existing somewhere in the mind; it is a set of epistemic and pragmatic actions with different but connected goals and temporalities. It is essential to grasp this circular structure. Predictive error drives bodily movement. In turn, bodily movement—composed of a series of micro-actions—generates a stream of sensory data at multiple levels: interoceptive, exteroceptive, and proprioceptive. Perception, cognition, and action work closely together "to minimize sensory prediction errors by selectively sampling, and actively sculpting, the stimulus array. They thus conspire to move a creature through time and space in ways that fulfil an ever-changing and deeply inter-animating set of (sub-personal) expectations" (Clark 2013, 186).

1.4 THE DARK ROOM PROBLEM

A classic objection to PP is as follows: if the goal of every cognitive activity is to reduce error and uncertainty related to predictions, then human behavior should be essentially conservative. We would tend to create increasingly predictable and boring environments to retreat into until death. Yet this is not the case at all: humans seek novelty and are inherently curious. Our brains not only tend to explore the environment, seeking new situations, but a certain level of uncertainty even seems to generate pleasure. How is this possible? This is what Friston et al. (2012b) refer to as the "dark room problem" and what Clark (2023) calls the "simple death trap." One straightforward answer might be this: The environment in which living organisms operate is highly uncertain, which also necessitates confronting uncertainty by creating controlled bubbles of uncertainty to gather more information and develop more effective solutions to specific problems. Moreover, living organisms have basic needs such as eating, drinking, and reproducing, and these necessitate some exploratory behavior (Friston et al. 2012b). In other words, we accept a certain level of uncertainty to achieve our goals and thus minimize uncertainty in the long term. Staying in a highly predictable "dark room" would lead to physical dissipation and depression.

Another answer refers more to learning and the plasticity of adaptation. In a highly complex environment, learning is the fundamental key to long-term success. Naturally, learning involves discovering new things, which means embracing a certain degree of uncertainty—"a kind of controlled uncertainty" (Clark 2023, 94). In this context, the central goal is to find a balance between situations that are too predictable (where there is nothing new to learn) and those that are too unpredictable (where there is too much to learn). This is what we can call the trade-off between exploration and exploitation. Also, the cultural and social world is structured in such a way as to promote a certain level of exploration and uncertainty—pursuing novelty as a strategy to foster greater adaptation.

A more complex solution is provided by Clark (2017, 2023) through the notion of "interoceptive predictive coding," which refers to the flow of predictions concerning the internal states of the body. This set of predictions about the body's internal physiological states overlaps with the set of predictions about states of affairs in the external world, that is, exteroceptive predictions. When, due to this overlap, the internal state of the body and the predictions about it change, the "salience" (i.e., expected information gain) we receive from the outside world also changes, and this can influence the overall process of precision-weighting. As Clark (2017, 526) states, "as your bodily states alter, the salience of various worldly opportunities alters too. Such estimations of salience are written deep into the heart of the predictive processing model, where they appear as alterations to the weighting (the 'precision') of specific prediction error signals." In other words, the overlap of prediction systems ensures that changes in interoceptive predictions cause changes in exteroceptive predictions, thereby altering precision estimations, and this provokes curiosity, that is, the search for new information and new situations more closely aligned with the body's needs. From this asymmetry of predictive systems, new predictions and precision estimations emerge, allowing the brain to dynamically adjust its focus and priorities based on the interplay between internal and external states. (For a critique of these positions, see Klein 2018. I do not pursue that debate here.)

1.5 CONCLUSION

The goal of this chapter was to provide an initial introduction to the FEP and active inference via PP. Of course, there are many other topics and challenges to explore (e.g., the relationship between PP and various psychopathologies or forms of neurodivergence). However, I chose to keep the presentation at a nonspecialist level to allow for a gradual entry into the FEP. As noted, PP should not be entirely equated with active inference. When we refer to PP, we mean a family of theories that sometimes explicitly draw on active inference concepts and, at other times, adopt different frameworks.

2

ACTIVE INFERENCE: A GLOBAL PICTURE

Despite the wealth of empirical data in neuroscience, there are relatively few global theories about how the brain works. A recently proposed free-energy principle for adaptive systems tries to provide a unified account of action, perception and learning.
—K. Friston, "The Free-Energy Principle: A Unified Brain Theory?" (2010a)

Over the last fifteen years, an ambitious explanatory framework has been proposed to unify explanations across biology and cognitive science. Active inference, whose most famous tenet is the free energy principle, has inspired excitement and confusion in equal measure.
—S. Mann, R. Pain, and M. Kirchhoff, "Free Energy: A User's Guide" (2022)

Responding to philosophical deconstructions of the FEP has now become a familiar part of my monthly routine.
—K. Friston, "The Ultimate Trick? Comment on: 'The Markov Blanket Trick: On the Scope of the Free Energy Principle and Active Inference' by Raja et al." (2022)

Once you start looking at the world through a Bayesian lens, you do start seeing Bayes' theorem everywhere.
—T. Chivers, *Everything Is Predictable* (2025)

What is active inference? Is it just deep learning applied to the brain? Active inference aims to provide a unified account of behavior and cognition in sentient organisms—and, more broadly, in any self-organizing system that maintains its physical integrity over time. How do living organisms maintain their unity and persistence in a highly uncertain environment? What is the role of learning in shaping behavior and perception? How can complex adaptive behaviors emerge from simple principles? The main thesis of active inference is that "living organisms can only maintain their bodily integrity by exerting adaptive *control* over the action-perception loop" (Parr et al. 2022, 3) and they do that by minimizing surprise, or—better—a proxy of it, that is, what we call (variational or expected) free energy. In other terms,

> Active inference is a normative framework to characterize Bayes-optimal behavior and cognition in living organisms. Its normative character is evinced in the idea that *all facets of behavior and cognition in living organisms* follow a unique imperative: minimizing the surprise of their sensory observations. Surprise has to be interpreted in a technical sense: it measures *how much an agent's current sensory observations differ from its preferred sensory observations*—that is, those that preserve its integrity (e.g., for a fish, being in the water). Importantly, minimizing surprise is not something that can be done by passively observing the environment: rather, agents must adaptively control their action-perception loops to solicit desired sensory observations. This is the active bit of active inference. (Parr et al. 2022, 6, my emphasis)

Active inference casts all forms of purposeful behavior as ultimately Bayesian processes. Control is described as an inferential process. This brings active inference closer to other approaches such as planning-as-inferring (Attias 2003) or control-as-inferring (Levine 2018). However, active inference differs from these approaches in that control and planning are performed with variational inference, aiming to minimize the free energy. Moreover, active inference also differs from dual process theories, which tend to separate two systems, one more intuitive and faster, and the other slower and reflective (Kahneman 2011). For active inference, intuition and reflection, practical and theoretical, habitual and deliberative aspects of behavior are expressions of a single inferential and embodied process.

Now, the term "active inference" includes two basic concepts:

- *The complementarity of action and perception*. Action and perception are different ways of applying FEP, that is, two different but complementary ways to minimize surprise. Both are active processes. This is an idea we already saw in the previous chapter on PP. The brain is not an organ that passively assimilates data from the external world and then detects recurring patterns in it as in the traditional cognitivist model. Instead, the brain is an active organ, constantly making probabilistic predictions about its environment, in the sense that it constantly anticipates the flow of data, only to find the divergence between its own anticipations and the data itself.

- *Bayesian inference*. Action and perception are two forms of the same probabilistic reasoning. They are not only related, but also part of a unique form of *embodied rationality*—the brain is embodied, it is the whole body that reasons. This embodied rationality can be described as Bayesian inference, that is, a method of probabilistic inference in which Bayes' theorem is used to update the probability for a hypothesis as more evidence becomes available (Chivers 2025). It represents a way of thinking about how the probability of a theory or hypothesis should be adjusted in light of new evidence, allowing for a systematic way to incorporate prior knowledge along with new data in the process of statistical analysis. More technically, Bayesian inference is "a statistical procedure that describes the optimal way to update one's beliefs (understood as probability distributions) when making new observations (i.e., receiving new sensory input) based on the rules of probability. Specifically, beliefs are updated in light of new observations using Bayes' theorem" (Smith et al. 2022, 3). According to the Bayesian brain hypothesis, the brain processes information and makes decisions based on principles of Bayesian inference. A crucial element of active inference is the partial revision and extension of this thesis, showing that *optimal* Bayesian inference is impossible for living organisms, which instead produce *approximate* Bayesian inference based on the FEP. (For the Bayesian brain approach, see Doya 2007.)

In this chapter, I aim to expand on these two foundational points and examine their primary implications to provide a clear—though not exhaustive—overview of active inference and the FEP.

2.1 A NORMATIVE APPROACH

Active inference is a normative approach; it aims to provide a coherent explanation of how a living organism *should* behave in a complex environment. The normative dimension here does not have a moral or ethical character. Instead, it means that cognitive activities and behavior in living organisms are evaluated in relation to an essential parameter: the minimization of uncertainty, or its computationally tractable proxy, free energy. Active inference models are normative in the same way that statistical or classical physics models are normative. Namely, they provide generalized, idealized, and simplified descriptions of the behavior of objects or systems.

Since the FEP is normative, it outlines how organisms should behave in order to protect their structural integrity; therefore, the organisms we observe successfully existing can be expected to behave in accordance with the FEP. In other words, the FEP asserts that the most probable (actual and future) behavior of a dynamical system is the one that minimizes free energy the most. Thus, to persist in a highly uncertain environment, an active inference agent is likely to choose the sequence of actions that best minimizes free energy.

2.2 SELF-ORGANIZATION

The FEP is closely connected to the concept of a self-organizing system. This latter was formalized in the field of cybernetics (Foerster 2003) and later expanded to various disciplines, including physics (Haken 1983), biology (Bell & Deater-Deckard 2007), and neuroscience (Friston 2013; Friston et al. 2011; Friston et al. 2017). (On the history of the concept of self-organization, which begins long before cybernetics, see Keller 2008.) Self-organization can be generally defined as the spontaneous emergence of spatiotemporal order or patterns in physical and biological systems, arising from the interactions of their components with the environment. In this process, the properties of a larger, higher-scale system emerge from—and depend on—the interactions of its smaller-scale components. For instance, when wind blows over a uniform sand surface, regularly spaced ridges form as gravity and wind speed influence the movement of sand particles (Ciaunica et al. 2023). In turn, the sand's surface alters the airflow, further shaping the ripples.

However, according to active inference scholars, biological organization extends beyond mere emergent complexity. It fundamentally consists of nested, goal-seeking agents—ranging from molecular pathways to entire organ systems—that engage in homeostatic and allostatic processes. These agents not only exhibit complex behaviors but also act with specific purposes, such as minimizing or maximizing certain variables and solving problems by navigating various spaces, including transcriptional, physiological, and anatomical spaces, in addition to the familiar behavioral space. For example, human bodies are biological systems that "share—with all other physical systems—the property of being instantiated in time and space (e.g., our body occupies a given position and volume in space at a given time)" (Ciaunica et al. 2023, 4). Human bodies "tend to avoid entropic dissipation by engaging in self-organization, with the (apparent or manifest) goal of maintaining their internal and sensory states within certain bounds, thereby evincing a homeostatic balance" (Ciaunica et al. 2023, 4). Active inference and the FEP are closely connected to the concept of self-organization because they both provide a framework to explain how systems—biological, cognitive, or even artificial—maintain their structure (homeostatic balance) and function despite external perturbations.

With this foundation in place, the following sections delve into some of the core elements of the most traditional formulation of the FEP: the notion of systems in a nonequilibrium steady state (NESS), the Markov blanket formalism, and the concept of variational free energy (figure 2.1). A crucial point I will underline is that certain aspects of this traditional formulation have recently been superseded in the literature.

2.3 NONEQUILIBRIUM STEADY STATES

How does an organism minimize uncertainty? The first step in addressing this question is to clarify what we are talking about—that is, to identify the subject of our discussion. For anything to count as a "thing," it must be distinguishable from everything else, including its surrounding environment. In the classical formulations of the FEP, scholars approached this question by first introducing the concept of a nonequilibrium steady state (NESS). "We define a 'system' (according to the FEP) as a dynamical system that possesses a NESS, which it maintains over a significant period, and whose dynamics are structured to satisfy the 'Markov blanket condition'" (Millidge et al. 2021, 5). Importantly, the NESS assumption is adopted in the traditional formulations of the FEP purely for convenience—it makes the math far more tractable, but it is not a fundamental requirement (Ramstead et al. 2023).

A nonequilibrium steady state describes a system that, despite ongoing interactions with its environment, maintains a statistically stable probability distribution over time. While individual states fluctuate, the overall distribution remains approximately constant, maintaining a NESS rather than reaching thermodynamic equilibrium. Crucially, this implies that the

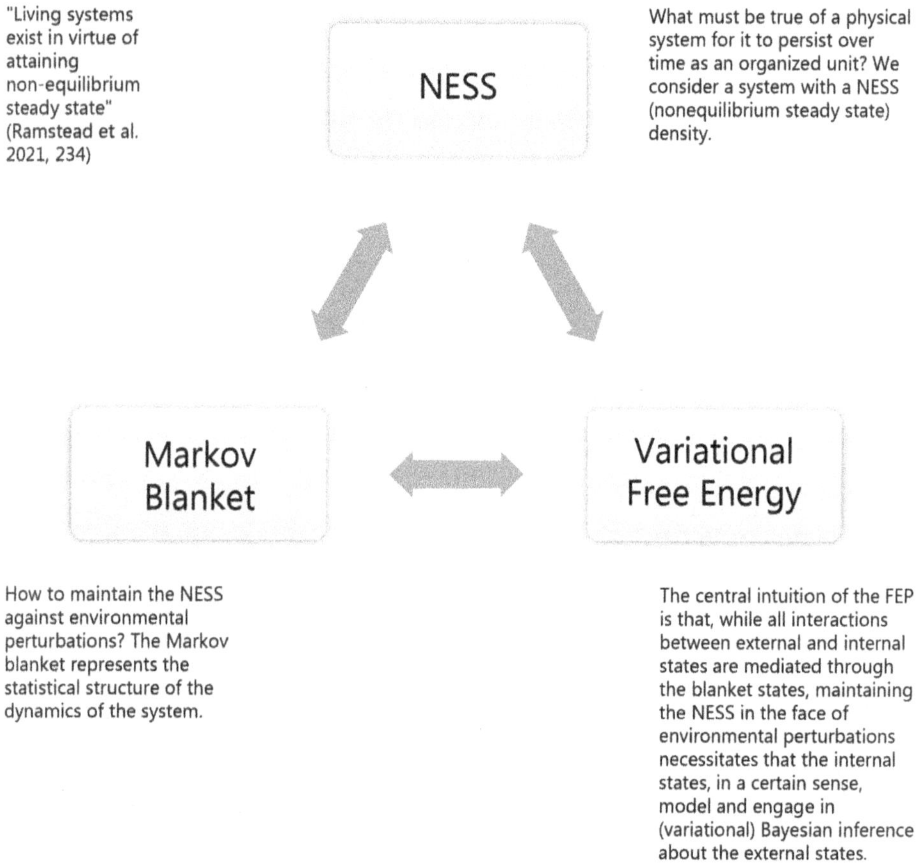

Figure 2.1
The conceptual architecture of the FEP. A system in a nonequilibrium steady state (NESS) sustains itself far from thermodynamic equilibrium by exchanging energy and information. Using Markov blankets and variational free energy, it performs approximate Bayesian inference. Bidirectional arrows represent a relationship where both concepts influence each other. This model obviously reflects the classic view of the FEP. As I try to show in this chapter, in recent years the literature on FEP has made significant progress, essentially demonstrating the independence of the FEP from the assumption of a NESS.

system does not passively equilibrate but instead sustains its statistical structure through continuous energy dissipation and exchange with the environment (for a more technical analysis, see Friston 2019, 10–13; see also Nicolis & Prigogine 1977; for a general introduction to this topic, see Prigogine 1996).

In thermodynamic equilibrium, a system reaches a state where macroscopic properties (such as temperature, pressure, and particle distribution) remain constant over time due to the absence of net energy and matter flows. The system's probability distribution over microstates is time-invariant: microscopic fluctuations still occur, but the overall statistical distribution remains stable. A classic example is a sealed gas in a container that eventually achieves a uniform distribution, after which no further large-scale changes take place. In contrast, a system in a NESS maintains a statistically stable distribution over time despite continuous exchanges of energy and matter with its environment. It persistently occupies a set of preferred states, even while interacting dynamically with its surroundings. Schrödinger (1944), in *What Is Life?*, famously described living systems as those that avoid thermodynamic decay by importing "negative entropy" (negentropy) from their surroundings—extracting order to maintain internal structure. This insight foreshadowed modern views of self-organizing nonequilibrium systems in biology and cognition.

A foundational axiom of the traditional formulations of the FEP is as follows: "We assume a system with a NESS density to which it can converge and focus on understanding the system's dynamical behavior at this density" (Millidge et al. 2021, 5). For a system to maintain its integrity over time, it must possess a stable NESS density, to which it continually converges despite external perturbations. The traditional formulations of the FEP assume this density exists and seeks to characterize the system's behavior in terms of minimizing free energy *relative to this density*.

This can be explained more simply by highlighting two key points:

1. *Entropy and uncertainty*. In physics, entropy quantifies, roughly, how many microscopic configurations (microstates) are compatible with a system's macroscopic properties. A system with higher entropy has more possible states (or the same number of states with a more even probability distribution), making it harder to predict its specific state. In information theory (Shannon 1948), entropy measures uncertainty or surprise, representing

the minimum amount of information required to describe a system. As Solms succinctly puts it: "The more yes/no questions one needs to answer to describe a system, the greater its entropy" (2021, 157). While physical entropy and informational entropy share a mathematical foundation, they describe different phenomena: physical entropy relates to energy distribution and disorder in thermodynamic systems, whereas informational entropy pertains to uncertainty and the efficiency of data representation. Both reflect uncertainty, but neither necessarily implies an increase in useful or meaningful information.

2. *Living systems and the management of entropy.* Living systems operate within the constraints of the second law of thermodynamics, maintaining low-entropy internal organization by actively managing entropy. They achieve this by limiting the range of states they can occupy, favoring a subset of states critical for survival. These preferred states are shaped by evolutionary processes and environmental conditions. Living systems do not resist entropy outright but reduce internal uncertainty by managing it through regulatory strategies and predictive models. These models help anticipate environmental changes and minimize the need to process unpredictable new information. In terms of information theory, this corresponds to decreasing the entropy associated with vital states, thereby optimizing the system's ability to maintain control over its environment.

Some critics have noted that theoretical approaches in the life sciences may conflict with the central assumption of the NESS (Colombo & Palacios 2021). In response to such concerns, the recent literature on the FEP and active inference—from essentially Friston et al. (2023a), Friston et al. (2023b), Ramstead et al. (2023), and Sakthivadivel (2022a) to Beck and Ramstead (2025)—has significantly nuanced the connection between the FEP, the NESS, and the second law of thermodynamics by developing a new formulation of the FEP. This latter shifts from a state-based formulation to a path-based one that no longer requires the assumption of a NESS. In some of the more advanced FEP formulations, it is shown how the FEP can be derived simply from Jaynes' principle of maximum entropy or maximum caliber under the constraint that there are boundaries of certain kinds (Beck & Ramstead 2025). These formulations show that, under certain assumptions, the FEP can be recovered from maximum caliber plus explicit boundaries. Whereas the classical formulations (Friston 2019) start from the equations of statistical physics (the Langevin or Fokker–Planck equation) and arrive at an equivalent description of the system in information-theoretic terms (a gradient descent over a free-energy functional), the more advanced approaches instead begin with maximum caliber to show how, once boundaries are imposed, it becomes the FEP—from this point of view, FEP systems maximize entropy, under the constraint that their boundaries exist and have such-and-such statistics.[1] This aims to address the criticism that the FEP depends on unjustified and unrealistic presuppositions (e.g., the NESS).

2.4 MARKOV BLANKETS

Let us now move on to the second element of our conceptual architecture. The concept of Markov blanket is crucial for defining and understanding the dynamic structure of active inference agents. These latter self-organize via Markov blankets—and Markov blankets of Markov blankets (Friston 2019; Kirchhoff et al. 2018; Parr et al. 2019). A Markov blanket defines the boundaries of a system in a statistical sense. It introduces a partition of states that is fundamental for the FEP.

As mentioned, according to the FEP, a dynamical system tends to remain in its preferred states by minimizing variational free energy *with respect to its internal states*, thereby *maintaining an upper bound on sensory surprise*. In the classical formulations of the FEP, this is analyzed under the assumption that the system has a NESS. Free-energy minimization explains why it remains near that NESS, not how it maintains it in a causal sense. The preferred states, known as "attracting states" (Corcoran et al. 2020, 5), form the "invariant set towards which the system inevitably evolves over time," shaped by evolutionary and adaptive processes within its environment. An example is the behavior of desert plants, such as cacti, which must conserve water to survive in arid environments. Through specialized adaptations like thick, waxy cuticles and stomata that open only at night, these plants maintain internal moisture levels and achieve *homeostasis*. Homeostasis is the ability to sustain a stable internal environment despite external fluctuations.

The term "homeostasis" was coined by the American physiologist Walter Bradford Cannon in 1929, building on earlier ideas about the human body's self-regulation mechanisms. Homeostasis involves various physiological processes that ensure key variables—such as body temperature, pH, glucose levels, and electrolyte balance—remain within narrow, optimal ranges. For instance, the human body maintains a relatively constant core temperature of about 37°C (98.6°F) through mechanisms like sweating, shivering, and blood flow adjustments, regardless of the external temperature. The idea of homeostasis is centered on the notion that living organisms have set points or ranges of acceptable values for various physiological variables. When these variables deviate from the set points, feedback mechanisms are activated to bring them back to their optimal levels. For example, if blood glucose levels rise after a meal, the pancreas secretes insulin to lower the glucose level back to the normal range (Cummins 2014).

However, homeostasis alone is not sufficient to explain the complex dynamics of adaptation. It is also connected to dynamics that researchers refer to as *allostasis* (Schulkin & Sterling 2019). The term "allostasis" was first introduced by Peter Sterling and Joseph Eyer in 1988. Unlike homeostasis, which emphasizes maintaining stability through a fixed set point, allostasis focuses on the dynamic process of achieving stability through change. It involves the adjustment of physiological systems to meet anticipated demands or stressors, often in a more flexible and adaptive way than the rigid set-point model of homeostasis. It is the process by which the body anticipates and responds to stressors by altering the set points of various physiological systems. This could involve changes in heart rate, hormone levels, or immune function, depending on the specific demands of a situation. For instance, during a stressful event, the body might increase blood pressure and release stress hormones such as cortisol to prepare for a "fight or flight" response. After the stressor is removed, the body readjusts these systems back to their baseline levels. Active inference scholars (Pezzulo et al. 2015) have attempted to reinterpret allostasis through the lens of the FEP, essentially adapting it to align with homeostasis and the imperative to minimize free energy. However, as Nave (2025) aptly points out, integrating allostasis into homeostasis within the FEP framework poses challenges (further discussion on this will follow).

Now, a fundamental prerequisite for the homeostasis/allostasis cycle is the ability of a system to establish a boundary between itself and its surrounding environment—this is also the philosophical premise of Friston (2019, 4), for which "a 'thing' is distinguishable from something else and from no 'thing.'" This boundary ensures not only a certain degree of autonomy but also a necessary level of interaction with the environment. One way to formalize this condition is by using the concept of the Markov blanket, that is, "a statistical partitioning of a system into internal states and external states, where the blanket itself consists of the states that separate the two" (Kirchhoff et al. 2018, 1; Friston 2019, 16–18; Parr et al. 2019). The Markov blanket divides the system into three groups of statistical variables: internal states, external states, and blanket states. The blanket states are further divided into active states and sensory states. In this context, states stand in for any variable that locates the system at a particular point in the states space. In other words, the core philosophical idea is that "thingness" depends on sparseness—that is, on which variables are not directly connected. Crucially, the statistical boundaries defined by the Markov blanket do not necessarily correspond to material boundaries—and this also allows the Markov blanket concept to be applied to events, rather than just physical things or objects (Kirchhoff et al. 2018).

This structure ensures that internal and external states remain independent while being connected through the active and sensory states. Active and sensory states "shield" the internal states by creating a statistical boundary. This means that if we know the sensory and active states, the external states will not provide any additional information about the internal states, and the same holds in reverse. Therefore, we have a kind of circular or cyclical causality: external states cause sensory states that influence—but are not influenced by—internal states, while internal states cause active states that influence—but are not influenced by—external states. Put simply, internal states cannot directly affect external states but can do so indirectly by influencing active states. Likewise, external states cannot directly impact internal states but can do so indirectly by affecting sensory variables. As Friston (2019, 17) claims, "the dependencies induced by Markov blankets create a circular causality that is reminiscent of the action-perception cycle." Cyclical causality here means that "external states cause changes in internal states, via sensory states, while the internal states couple back to the external states through active states, such that internal and external states influence each other in a vicarious and reciprocal fashion" (Friston 2019, 17). Consequently, the internal and external states tend to synchronize over time, much like two pendulums attached to opposite ends of a wooden beam gradually swinging in unison (Kirchhoff et al. 2018).

The Markov blanket structures and governs the process of active inference and the FEP—according to Solms (2021, 166), the presence of a Markov blanket is the essential precondition for subjectivity or selfhood in a self-organizing organism, even in the absence of consciousness; one could even hypothesize that what we call personal identity or subjectivity is nothing more than a highly complex, temporally extended hierarchy of nested Markov blankets. First introduced by Judea Pearl as a statistical concept (Pearl 1988), the Markov blanket is a fundamental notion in probabilistic graphical models, particularly in Bayesian networks, which are widely used for representing the dependencies among a set of random variables. In the context of a graphical model, the Markov blanket of a particular node (or variable) encapsulates all the information needed to predict that node, shielding it from the rest of the network.

Technically, the Markov blanket comprises three distinct categories of variables:

- *Parents:* Variables that directly influence the target variable. These are causal precursors, determining the likelihood of the target variable's states.
- *Children:* Variables directly influenced by the target variable. These represent the consequences of the target variable's states.
- *Spouses:* Variables that influence the target variable indirectly by acting on its children (more precisely, spouses carry more information about the target variable in conjunction with parents and children). These variables mediate second-order dependencies, completing the causal structure around the target variable.

Together, these components form a closed dependency set that fully accounts for all statistical and causal influences on the target variable. For example, using the dependency graph shown in figure 2.2, we can say that the Markov blanket of whether the sprinkler is operating or not comprises the season (parents), wet grass (children), and whether it rained last night or not (spouse or coparent):

Any additional information from outside this Markov blanket is redundant because it does not improve the predictability of the target variable (Kirchhoff et al. 2018; Ororbia & Friston 2024; Parr et al. 2019; Parr et al. 2022).

It is important to note that in the path-based formulation of the FEP (Friston et al. 2023a; Ramstead et al. 2023), systems—or "particles"—are classified along two dimensions that reflect their Markov blanket structure: dissipative versus conservative, and inert versus active. Dissipative particles follow noisy, entropy-producing trajectories that deviate from Hamiltonian (least-action) flow, whereas conservative particles approximately follow Hamiltonian dynamics with relatively small fluctuations. In this formulation, the

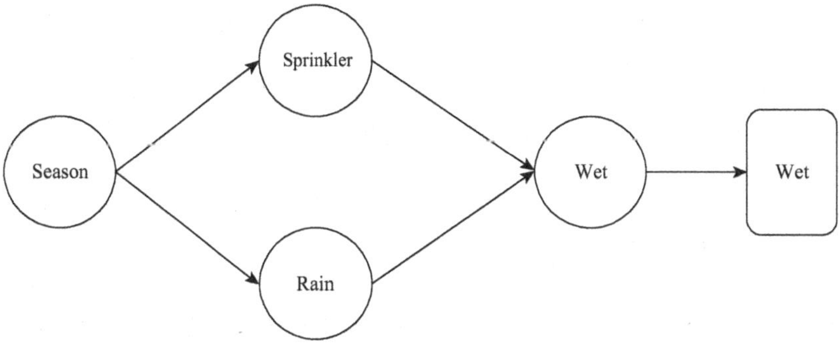

Figure 2.2
Markov blanket of the sprinkler state: its parent (season), its child (wet grass), and its co-parent/spouse (rain last night). Conditioning on these nodes makes the sprinkler independent of the rest of the network.

FEP is cast as a stochastic variational principle in which path probabilities are governed by a free-energy functional that plays a role analogous to an action functional (Friston et al. 2023a, 42). Inert particles lack active states and therefore minimize free energy only through perceptual inference, by updating internal states in response to sensory input. Active particles include active states that couple internal beliefs to environmental dynamics, enabling free-energy minimization through both belief updating and changes in active states expected to realize preferred outcomes. Among active particles, Friston et al. (2023a) distinguish ordinary from strange particles. Ordinary particles explicitly represent actions as control variables, whereas strange particles treat their active-state trajectories as latent paths that must be inferred from sensory consequences, alongside external states. In this latter case, both environmental and active trajectories are modeled as hidden causes of sensation, such that the particle must infer its own actions—an architecture that can be interpreted as enabling more autonomous, agent-like behavior. Friston et al. 2023a and Ramstead et al. 2023 offer a more physical interpretation of the FEP, grounded in the study of dynamical systems, compared to a more cognitive science-oriented interpretation such as Parr et al. 2022. These two approaches are different but complementary. Alongside these two interpretative lines, there is also a third one, the quantum interpretation: see Fields et al. 2022. I do not discuss this latter in the present book.)

From a philosophical point of view, Bruineberg et al. (2022) draw attention to a persistent confusion in the literature surrounding the use of

Markov blankets. They identify a misunderstanding between the formal, epistemic role of Markov blankets as tools for Bayesian inference and their metaphysical application within the FEP. To address this, the authors propose a conceptual distinction: "Pearl blankets" to denote the traditional epistemic application and "Friston blankets" to represent the metaphysical construct introduced in the FEP framework. Building on this distinction, they critically evaluate claims that apply Markov blankets to philosophical problems. They argue that the literature should clearly differentiate between two distinct research programs: "inference with a model" and "inference within a model." While only the latter can perform metaphysical work using Markov blankets, it requires additional assumptions and cannot be justified solely by the success of the mathematical framework. (More on this later.)

Beyond these philosophical considerations, it is worth noting that the concept of a Markov blanket is not just theoretical; it has several practical implications and applications across different fields (for technical details, see Bishop 2012; Whittaker 2009). Across these applications, the core idea remains the same: identifying a set of nodes (variables) that renders a given node conditionally independent of the rest of the network. In more complex models, such as dynamic Bayesian networks that extend Bayesian networks to temporal data, the notion of a Markov blanket is adapted to account for dependencies that evolve over time. Similarly, in hierarchical models, where variables are organized in layers or groups, the Markov blanket is defined with respect to the conditional independence structure induced by that hierarchy. However, despite its utility, identifying a Markov blanket is not always straightforward, especially in networks with complex dependencies or hidden variables. In some cases, the Markov blanket may be quite large and, in certain approximate or learned models, alternative sets that screen off a node from the rest of the network may also exist. Moreover, the assumptions underlying the Markov blanket—such as the network structure and the independence relations encoded by the graph—may not hold perfectly in real-world data. These challenges have motivated ongoing research on refining the concept, improving algorithms for Markov blanket discovery, and extending it to more complex models.

2.5 BAYESIAN INFERENCE

In this section, I introduce key notions for understanding the third element of our conceptual architecture (figure 2.1): Bayesian inference. This foundational knowledge is crucial for approaching active inference and the FEP. Indeed, the dynamics of persistent, bounded systems may be conceptualized as inferential processes. Specifically, the states within the Markov blanket seem to infer the states outside it. In other words, the existence of a Markov blanket implies that internal states evolve so as to minimize variational free energy conditioned on sensory states. By doing so, the system maintains its functional and structural integrity and resists dissipation, effectively inferring the hidden causes of its sensory signals. (Friston 2008, 2010a, 2010b, 2013, 2019; Friston et al. 2006).

The purpose of Bayesian inference is to find the best hypothesis for what you see. Given that the lawn is wet, what is most likely? That it rained all night? That you left the watering system running? What is the best explanation, the most likely cause of the observed data? As Seth (2021, 97) claims: "Bayesian reasoning is all about reasoning with probabilities. More specifically, it is about how to make optimal inferences—what we've been calling 'best guesses'—under conditions of uncertainty. 'Inference' . . . just means reaching conclusions on the basis of evidence and reason." Bayes' rule is "a mathematical recipe for going from what we already know (the prior) to what we should believe next (the posterior), based on what we are learning now (the likelihood)" (Seth 2021, 98).

Bayesian inference is a form of probabilistic inference based on Bayes' theorem (Evans & Rosenthal 2009, ch. 7). It involves updating beliefs about the state of the world (encoded as probabilities) in light of new evidence. Specifically, it provides a mathematical framework for combining prior knowledge (prior probabilities) with observed data (likelihood) to produce updated knowledge (posterior probabilities). This process starts with prior probabilities, representing initial beliefs about a hypothesis, and adjusts these beliefs in light of new evidence, resulting in a posterior probability. By continuously updating probabilities as more data becomes available, Bayesian inference allows for dynamic and flexible reasoning, making it particularly useful in fields such as machine learning, statistics, and various scientific disciplines. In many situations, the process of updating of Bayesian best guesses with new data

> happens over and over again, in an endless cycle of inference. On each iteration, the previous posterior becomes the new prior. This new prior is then used to interpret the next round of data to form a new posterior—a new best guess—and the cycle repeats. If your lawn is wet two mornings in a row, your best guess about the cause on the second day should be informed by your best guess on the first day, and so on as each new day comes along. (Seth 2021, 100)

From the active inference perspective, every dynamical system, in order to persist, formulates hypotheses about the world—expectations about reality—and tests them to gradually refine these models. This means that a system must rely on models of the world—but these models are never static; they require continuous refinement and adaptation. Therefore, identifying and understanding error is crucial.

2.6 MACHINERY 1: CAUSAL MODELS OF THE WORLD

To perform Bayesian inferences, a system must be equipped with a generative model (we have already anticipated this concept informally in chapter 1), which is a probabilistic and causal model of the world. The generative model consists of a set of hypotheses or predictions (represented as probability distributions) about the hidden causes (referred to as hidden states, because they are not directly observable) of sensory data that the brain receives and processes (which, in contrast, are observable and directly measurable).

As Dayan and Abbott (2001, ch. 10) clearly explain, a generative model is, quite literally, a model designed *to generate synthetic data from assumed causes*. The only information available to the system (e.g., the brain) consists of probability distributions over possible causes, and its task is to infer the causes of the observed data. To accomplish this, the system constructs simulations of the data, effectively building a generative model that can simulate and produce probability distributions mapping data to its underlying causes. Crucially, these synthetic probability distributions must closely resemble the actual distributions. This requires the system *to adjust the parameters of its synthetic distributions, progressively bringing them closer to the real ones*. However, because the system does not have direct access to the real distributions, it can rely on only one strategy: predicting (simulating) the data and minimizing prediction (simulation) errors. By tracking these errors, the system iteratively refines its parameters, gradually aligning its synthetic distributions with the real ones. This guessing and tweaking is what we will call *variational Bayesian inference*. There is also a more technical meaning of variational Bayesian inference: see MacKay (2003, 429), who calls it "variational Bayes."

A generative model is represented by a joint probability distribution $P(x, y)$, describing the relationship between observed data (y) and the hidden states (x) that generate them. This joint probability can be broken down into two parts: *the prior* $P(x)$, which is the hypothesis about the probability of x (e.g., the general probability of encountering a tiger in that forest at that time based on my previous experiences and general knowledge of forests), and *the likelihood* $P(y|x)$, which is a conditional probability, meaning the probability of y (the observations) given x (e.g., the probability of seeing the bush move given that there is a tiger behind it). The likelihood reflects the agent's knowledge of how observations can be generated by a hidden cause (if that is the cause, *then* these are the likely data). The likelihood is essentially a model of the world based on the probability of the prediction in question. These are the essential elements of the generative model.

A *probability distribution* is a mathematical function that describes the likelihood of various outcomes for a random variable. A random variable is a quantity that can vary due to chance, such as the probability that there is a tiger behind a bush. Probability distributions can be categorized as either discrete or continuous. For a discrete random variable, the distribution assigns a specific probability to each possible value, with all probabilities summing to 1. A classic example is rolling a die, where each face (1 through 6) has an equal probability of 1/6. For a continuous random variable, the distribution is represented by a probability density function (PDF), which integrates to 1 over the range of possible values. A typical example is the normal distribution, where the random variable can take any real value, and probabilities are determined by the area under the PDF curve over a given interval. In essence, a probability distribution for a random variable x defines the likelihood of x taking on specific values or falling within certain ranges.

A normal (Gaussian) distribution is fully characterized by two parameters: the mean and the precision (1/variance) (Bishop & Bishop 2025; Gelman et al. 1995). Together, these parameters define both the central tendency (mean) and the spread or confidence in that central estimate (precision):

- The *mean* corresponds to the most plausible value of a belief, whether derived from a prior, a likelihood, or a posterior distribution. For example, the posterior mean, calculated after updating the prior with new evidence, indicates the best estimate of a hidden state based on current data. Importantly, the mean does not capture variability but solely reflects the central value around which the belief is centered.

- *Precision*, defined as the reciprocal of variance, measures the confidence or reliability associated with the mean. Higher precision corresponds to a sharply concentrated distribution, indicating greater certainty and lower variability around the mean. Conversely, lower precision reflects a broader distribution, signifying greater uncertainty. In Bayesian inference, precision plays a critical role in weighting different sources of information—such as prior beliefs versus observed evidence—during belief updating. A highly precise prior, for example, can dominate weak evidence, while strong evidence can override a low-precision prior.

These two properties are interconnected and essential for interpreting a probability distribution. The mean provides a point estimate that represents the "best guess," while the precision quantifies the reliability of that estimate. Together, they enable Bayesian models to express not only what is believed but also how confident the model is in those beliefs, which is crucial for decision-making, prediction, and uncertainty quantification in complex systems. For more technical details and discussions, see Bishop and Bishop (2025), McElreath (2020), and Murphy (2012).

Now, *Bayes' rule* tells us how to update the initial prediction after receiving new data, thus moving from the prior $P(x)$ and the likelihood $P(y|x)$ to the *posterior probability* $P(x|y)$. Essentially, to perform the inference we must invert the likelihood. The following is the classic formula for Bayes' rule:

$$P(x|y) = \frac{P(y|x)P(x)}{P(y)} \tag{2.1}$$

In a nutshell, Bayes' rule provides "a very precise measure of how much a new piece of evidence should make us change our ideas about the world" (Frith 2007, 123). It is a fundamental result derived from two core principles of probability theory: the product rule and the sum rule. These principles not only provide the foundation for Bayes' rule but also enable the concept of *marginalization*, a critical operation for deriving probabilities over subsets of variables in joint distributions. Marginalization allows us to compute the marginal probability of a variable y by integrating out or summing over other variables in the model (e.g., x). This operation is essential for evaluating a model's predictive power, as it helps determine how well the model accounts for observed data. This measure, known as *model evidence*, represents the probability of the data $P(y)$ under the given model and serves as a key metric for assessing the model's ability to predict new events and generalize effectively.

Let us clarify further with another example. Imagine a person sees a bush move. Based on their previous experiences and general knowledge of the environment, this person forms two hypotheses: there could be a rhinoceros or a frog behind the bush. The probability assigned to the presence of the rhinoceros is 0.9, while the probability of a frog being there is 0.1. This means that the person can also predict what observations they should expect if there is a rhinoceros or a frog behind the bush, and this is the likelihood. For instance, if the person sees something jump, then their prior will be wrong and will need to be updated. (For more details, see Parr et al. 2022; Smith et al. 2022.)

In short:

Priors Two hidden states
$P(x) = \text{rhinoceros } 0.9, \text{ frog } 0.1$

Likelihoods Beliefs about possible observations given that $P(x)$ is true
$P(\text{jump} \mid \text{rhinoceros}) = 0.1$
$P(\text{no jump} \mid \text{rhinoceros}) = 0.9$
$P(\text{jump} \mid \text{frog}) = 0.8$
$P(\text{no jump} \mid \text{frog}) = 0.2$

Now, to calculate the posteriors, we need the model evidence, which is $P(y)$. As mentioned, the operation of marginalization allows us to calculate the probability of a single event with respect to the model, that is, in relation to all the hypotheses within the model. In order to do this, we apply the following formula, known as the law of total probability:

$$P(\text{jump}) = P(\text{jump} \mid \text{rhinoceros}) \cdot P(\text{rhinoceros}) + P(\text{jump} \mid \text{frog}) \cdot P(\text{frog})$$

Substituting the given values:

$$P(\text{jump}) = (0.1 \cdot 0.9) + (0.8 \cdot 0.1)$$

$$P(\text{jump}) = 0.09 + 0.08$$
$$P(\text{jump}) = 0.17$$

In a nutshell, this value is the combined probability that a jump occurs considering both possible causes (rhinoceros or frog). The law of total probability, which relates marginal probabilities to conditional probabilities, helps to marginalize out the different scenarios (rhinoceros and frog) to find the overall probability of the event (jump) under that model, that is, all the hypotheses. Starting from this value, we can calculate the posteriors. So, after observing a jump, the updated probabilities are approximately:

For $P(\text{rhinoceros} \mid \text{jump})$

$$P(\text{rhinoceros} \mid \text{jump}) = \frac{P(\text{jump} \mid \text{rhinoceros})P(\text{rhinoceros})}{P(\text{jump})}$$

$$P(\text{rhinoceros} \mid \text{jump}) \approx 0.529$$

For $P(\text{frog} \mid \text{jump})$

$$P(\text{frog} \mid \text{jump}) = \frac{P(\text{jump} \mid \text{frog})P(\text{frog})}{P(\text{jump})}$$

$$P(\text{frog} \mid \text{jump}) \approx 0.471$$

Let us take it a step further. Logarithmic transformations of probabilities simplify calculations by converting multiplication and division into addition and subtraction, respectively. This approach is particularly useful in practice, as it avoids issues that arise when working with extremely small probabilities, which are common in complex, real-world scenarios. Importantly, log-probabilities can be converted back to standard probabilities using the exponential function (see Smith et al. 2022, appendix A). In active inference, negative log evidence is used as a measure of surprise—that is, the extent to which observed data deviates from what a model predicts. To calculate the negative log evidence, we begin with the log evidence, which is the logarithm of the model evidence. The model evidence quantifies how well a generative model explains the observed data, and its logarithmic form simplifies both computation and interpretation.

To illustrate, consider the marginal probability of observing an event $P(\text{jump}) = 0.17$. Taking the natural logarithm (base e; it can be denoted as $\log_e$ or simply ln):

$$\log_e(0.17) \approx -1.772$$

The negative log evidence is then:

$$-\log_e(0.17) \approx 1.772$$

This value measures how unexpected or surprising the observation is given the model. Specifically:

- High negative log evidence (significantly greater than 1) indicates a very surprising event, suggesting the model poorly predicted this outcome.
- Low negative log evidence (close to 0) indicates the event aligns closely with the model's predictions, implying minimal surprise.

In this example, the negative log evidence of approximately 1.772 indicates that observing a jump is modestly surprising given our priors and likelihoods. While a jump is not the most expected outcome, it remains well within the realm of possibility. This analysis highlights a key point: generative models can be compared based on their rate of surprise (negative log evidence) for a given set of observations. Models with lower surprise values are more effective at predicting data, often reflecting superior structure, priors, or variable definitions. Since, in active inference, generative models directly shape behavior, more effective models—those with lower surprise—enable more adaptive and accurate behaviors (Parr et al. 2022, 19). A fundamental note: negative log evidence measures how surprising a specific observation is under a model, while entropy measures the average surprise across all possible observations. In other words, negative log evidence is pointwise, and entropy is its expectation over the entire distribution.

Crucially, in active inference the generative model is not a content-bearing entity but rather a tool for action, used to select the best adaptive action policies. Therefore, active inference is always an *enactive* inference: "[It] is not merely a view of the brain as reducing the uncertainty of its sensory observations via perceptual inference. It concerns the active, selective sampling of the world by an embodied agent . . . active inference is the name of the formulation for policy selection" (Ramstead et al. 2019c, 225–226; Ramstead et al.

Table 2.1
Some essential terminological distinctions in the active inference framework

Concept	Definition	Key Features
Generative process	The actual causal mechanism in the external environment that produces sensory observations	• Real-world dynamics • Includes hidden states generating sensory data • Includes action-perception cycles in the environment
Generative model	A statistical model of the generative process that predicts and explains sensory data based on inferred hidden causes, enacted through adaptive behavior	• Specifies expectations about sensory inputs • Guides action by minimizing free energy • Exists as part of the organism's dynamics (not physically encoded)
Recognition model	An approximate posterior density over hidden states that represents the organism's "best guess" about the causes of its sensory input	• Embodied in the organism's internal states • Updated dynamically through active inference • Used for inference via model inversion

2019b). Generative models "are not explicitly encoded by physical states . . . it is the adaptive behavior of the system that implements or instantiates a generative model" (Ramstead et al. 2019c, 225–226). This means that the generative model must be distinguished from the *recognition model*. While the generative model generates predictions about the causes of sensory data (i.e., priors), which are actions (in other words, these predictions are the motives of the actions), the recognition model represents the posterior probability density over hidden states (it is called a recognition model because it allows one to recognize the most likely cause of a given observation). The recognition model, according to active inference, is physically encoded in the organism's internal states, such as neural activity, which embody the parameters required for inference. This physical encoding allows the system to dynamically update its beliefs based on sensory data by minimizing variational free energy with respect to its internal states. The generative model, by contrast, predicts how sensory inputs arise from hidden states, that is, the *generative process*. It defines the prior expectations about the relationship between hidden causes and observations, serving as a control framework for guiding actions and updating beliefs. However, the generative model *is not physically instantiated*; it is entailed by the system's adaptive behavior and enacted through its interactions with the environment. (For more accurate terminological distinctions, see table 2.1; see also Constant et al. 2019a.)

2.7 MACHINERY 2: NO OPTIMALITY

This is the crucial step that takes us from the Bayesian inference to active inference. As mentioned, this is not a technical exposition of the FEP. My aim is not to precisely define the mathematical structure of active inference (for this, see especially Bogacz 2017; Friston 2019; Millidge et al. 2021; Smith et al. 2022). Instead, my goal is only to identify some of its fundamental concepts and test their applicability in understanding technology and design. Nevertheless, I am convinced that without at least a working understanding of the mathematical formalism of the FEP, one cannot truly grasp its conceptual depth—which goes far beyond the simple mantra of uncertainty reduction.

In short, it is often impossible to strictly follow Bayes' rule. In many situations, the posterior and the model evidence cannot be calculated. This is due to two main interconnected reasons:

- The model evidence (and therefore the posteriors) is typically intractable to compute directly (except for the simplest cases) because it involves integrating or summing over all possible hidden states (Millidge et al. 2021, 11; Parr et al. 2022, 51–55; Smith et al. 2022, 5). This is computationally prohibitive due to the complexity of the models involved. More specifically, two interrelated issues can be distinguished: "The first is that for complex models, there may be many types of hidden states that all need marginalizing out, making the problem computationally intractable. The second is that the marginalization operation might require analytically intractable integrals" (Parr et al. 2022, 27).
- The time and energy resources of a living organism are limited. The active inference agent always operates in an environment that places limits on it.

This means, in other words, that we must reformulate Bayesian inference as variational Bayesian inference, framing it as an *optimization problem*. "Active inference appeals to a variational approximation of Bayesian inference that is tractable ['tractable' in terms of both computational complexity and bounded rationality]" (Parr et al. 2022, 64). One way to resolve this "is to convert this potentially difficult integration problem [the optimal Bayesian inference] into an optimization problem" (Parr et al. 2022, 64).

In other words, the problem becomes selecting a probability distribution (Q) that closely corresponds to the true posterior. To make this selection, we use the variational free energy (*F*). The approximate posterior that comes closest to the true posterior is the one that minimizes free energy the most (see figure 2.3). This approach builds on variational inference, a technique from machine learning used to approximate solutions to inference problems that are otherwise too complex to solve directly (see, for instance, Hinton & van Camp 1993). It is also important to note that while there are multiple approaches to approximate Bayesian inference, such as Markov chain Monte Carlo (MCMC) or Laplace approximation, active inference predominantly employs variational methods due to their computational efficiency and tractability in high-dimensional, continuous state, and observation spaces (Friston et al. 2006). Unlike MCMC, which requires costly sampling, variational inference reformulates inference as an optimization problem, making it more suitable for real-time applications in biological and artificial systems.

In short, variational free energy is a way to manage a fundamental quantity: *surprisal*, or negative log evidence ($-\ln P(y)$), that is, how unexpected the observation y is under a model—how badly the model predicts the data. High negative log evidence means that the observation y was highly unexpected, indicating a poor fit between the model and the data. As mentioned previously, evidence is the marginal probability of observation y under a model. The log evidence ($\ln P(y)$) is the natural logarithm of the evidence; it quantifies how probable the observation y is under a model, but transformed to a logarithmic scale. It is essential to note that both of these quantities—negative log evidence and log evidence—are always relative to a specific generative model, which is inherently subjective (the model may or may not correspond to the real generative process). The key point is that minimizing the negative log evidence is equivalent to maximizing the log evidence, thereby improving model fit. However, computing model evidence is generally intractable because it requires marginalizing over all hidden causes. As a result, we use a workaround: instead of directly minimizing surprisal, we minimize its proxy, known as variational free energy. This approach is technically identical to the evidence lower bound (ELBO) used in machine learning. In this context, variational free energy is the negative of the ELBO.

In this framework, we assume a true posterior distribution and minimize the divergence (measured as the Kullback–Leibler divergence; more on this later) between this true posterior and a tractable approximate posterior, which represents our best guess about the data. Minimizing this divergence implicitly reduces surprisal and, consequently, increases model evidence. Maximizing model evidence (i.e., confirming one's hypotheses about the world) by minimizing variational free energy is exactly what we call *self-evidencing* (Hohwy 2016).

Let me quote Friston (2009, 294; my emphasis):

> Because entropy is the long-term average of surprise, agents must avoid surprising states (e.g. a fish out of water). But there is a problem: *agents cannot evaluate surprise directly*; this would entail knowing all the hidden states of the world causing sensory input. However, an agent can avoid surprising exchanges with the world if it minimizes its free energy because free energy is always bigger than surprise.

Rather than directly calculating the true posterior distribution across the entire space of possible distributions—a computationally intractable task—a more practical approach is to approximate it using a simpler family of distributions defined by a small number of parameters. A common choice in Bayesian modeling of biological systems is the Gaussian distribution, as it provides computational tractability and aligns with empirical findings in neural coding and perception (Friston et al. 2006). Gaussian distributions are also widely used in variational inference and machine learning, where they simplify optimization problems and facilitate efficient computation.

Now, the approximate posterior, which is the recognition model, serves as a surrogate for the true posterior distribution. A technique to optimize the parameters of the approximate posterior is gradient descent, that is, an iterative method that adjusts parameters by calculating the gradient (rate of change) of an objective function and updating the parameters in the direction that reduces the function's value (Friston 2009, 295). Imagine you are hiking down a hill in thick fog—you cannot see the bottom, but you take small steps in the steepest downward direction, adjusting as you go, until you reach a low point. The objective function is like a map of the terrain, assigning a value to each position, where lower values indicate better

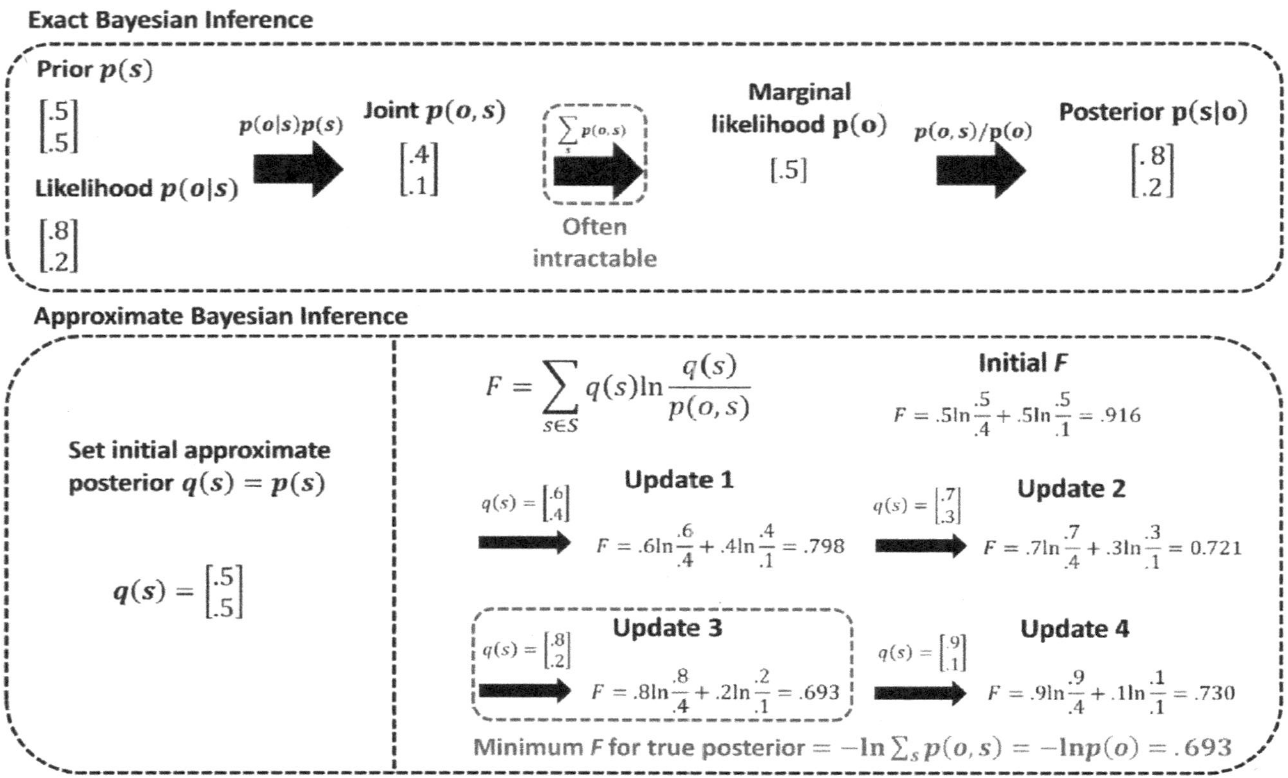
Exact Bayesian Inference
Prior $p(s)$
$\begin{bmatrix} .5 \\ .5 \end{bmatrix}$
Likelihood $p(o|s)$
$\begin{bmatrix} .8 \\ .2 \end{bmatrix}$
$p(o|s)p(s)$
Joint $p(o,s)$
$\begin{bmatrix} .4 \\ .1 \end{bmatrix}$
$\sum_s p(o,s)$
Often intractable
Marginal likelihood $p(o)$
$[.5]$
$p(o,s)/p(o)$
Posterior $p(s|o)$
$\begin{bmatrix} .8 \\ .2 \end{bmatrix}$
Approximate Bayesian Inference
Set initial approximate posterior $q(s) = p(s)$
$q(s) = \begin{bmatrix} .5 \\ .5 \end{bmatrix}$
$F = \sum_{s \in S} q(s) \ln \frac{q(s)}{p(o,s)}$
Initial F
$F = .5\ln\frac{.5}{.4} + .5\ln\frac{.5}{.1} = .916$
Update 1
$q(s) = \begin{bmatrix} .6 \\ .4 \end{bmatrix}$
$F = .6\ln\frac{.6}{.4} + .4\ln\frac{.4}{.1} = .798$
Update 2
$q(s) = \begin{bmatrix} .7 \\ .3 \end{bmatrix}$
$F = .7\ln\frac{.7}{.4} + .3\ln\frac{.3}{.1} = 0.721$
Update 3
$q(s) = \begin{bmatrix} .8 \\ .2 \end{bmatrix}$
$F = .8\ln\frac{.8}{.4} + .2\ln\frac{.2}{.1} = .693$
Update 4
$q(s) = \begin{bmatrix} .9 \\ .1 \end{bmatrix}$
$F = .9\ln\frac{.9}{.4} + .1\ln\frac{.1}{.1} = .730$
Minimum F for true posterior $= -\ln \sum_s p(o,s) = -\ln p(o) = .693$

Figure 2.3

This image (from Smith et al. 2022) captures the essence of variational free energy (VFE). Given x = states and y = observations, the goal is to find the approximate posterior $q(x)$ that best aligns with the true posterior $p(x|y)$ (Smith et al. 2022 use $s = x$ and $o = y$). To achieve this, we can only use the Kullback–Leibler divergence (D_{KL}), which represents the amount of information required to reconstruct $p(x|y)$ given full knowledge of $q(x)$. It is expressed as follows:

$$D_{KL}[q(x) \,||\, p(x \mid y)] = \sum_x q(x) \ln \frac{q(x)}{p(x \mid y)}$$

To find the approximate posterior $q(x)$ that best approximates the true posterior $p(x|y)$, we simply apply the gradient descent technique. Note that Smith et al. (2022) decompose the D_{KL} because this allows us to rewrite the VFE in a way that makes explicit the trade-off between accuracy and complexity. *Source:* R. Smith, K. Friston, and C. Whyte, 2022, “A Step-by-Step Tutorial on Active Inference and Its Application to Empirical Data,” *Journal of Mathematical Psychology* 107, p. 102632. https://doi.org/10.1016/j.jmp.2021.102632.

solutions. The goal of gradient descent is to find the lowest possible point, meaning the best solution given the function's constraints. Since gradient descent is used to minimize error, cost, or uncertainty, it seeks the lowest value on the map, though it may not always find the absolute lowest point (global minimum) if the landscape has multiple dips (local minima).

Crucially, in active inference, the optimization process balances two objectives: accuracy and complexity. Accuracy measures how well the approximate posterior explains the observed evidence, aiming to maximize this value to ensure the model aligns closely with the data. Complexity, on the other hand, quantifies how much the recognition model diverges from prior knowledge or assumptions. Minimizing complexity prevents the model from overfitting, which occurs when the model becomes too specific to individual observations and fails to generalize. These two objectives are typically combined into a single optimization criterion. As mentioned, this approach is structurally identical to the ELBO in machine learning (Bishop & Bishop 2025). The ELBO balances maximizing the fit to observed data with minimizing unnecessary deviations from prior regularities, effectively capturing the trade-off between overfitting and underfitting. By reformulating the problem in this way, the task of full Bayesian inference, which is otherwise computationally prohibitive, is reduced to a tractable and efficient optimization process that progressively aligns the recognition model with the true posterior distribution.

Let us take a step forward. Variational free energy operates on this very principle, that is, optimizing the parameters of the approximate distribution to achieve two primary goals:

- Ensuring the model accurately explains the data → **maximizing accuracy**.
- Avoiding an overly complex or overly specific model that fits only the current data → **minimizing complexity**.

This is understood as a trade-off between two key components:

- **Surprisal**, or **negative log evidence**. As mentioned, negative log evidence is commonly used to measure how surprising an observation is under a given model. The log evidence, $\log P(y)$, quantifies the likelihood of the observed data under the model in logarithmic terms. For probabilities $P(y)$, which range between 0 and 1, log evidence is always negative, reaching 0 only when $P(y) = 1$, which represents certainty. The negative log evidence, defined as $-\log P(y)$, is simply the negation of the log evidence. This transformation flips the sign, making negative log evidence a nonnegative (zero or positive) quantity. Negative log evidence plays a key role as a cost function in probabilistic modeling and machine learning. It evaluates the fit between the model and observed data. Smaller negative log evidence (closer to 0) indicates better model fit, as it corresponds to higher probabilities and less surprise. Larger negative log evidence reflects a poorer fit, corresponding to lower probabilities and greater surprise. Log evidence and negative log evidence are mathematical opposites. Log evidence, being the logarithm of probability, is a direct measure of the likelihood of the observed data in a logarithmic scale. Negative log evidence reverses the sign, converting a negative likelihood measure into a positive cost. This relationship ensures that the more probable (and less surprising) an event is, the smaller the negative log evidence will be, making it ideal for optimization tasks.

- The **Kullback–Leibler** (D_{KL}) divergence is an information-theoretic measure of disparity between probability distributions (Joyce 2011). It measures the relative entropy between two probability distributions, quantifying the amount of information needed to reconstruct the true posterior given the complete knowledge of the approximate. In other words, D_{KL} quantifies the dissimilarity between two distributions. It equals zero when the distributions are identical, and increases as the distributions diverge further (Smith et al. 2022, 7). It is not a true distance metric (Joyce 2011, 721). Bear in mind that in this context, information is quantified using a unit called nats, which is derived from the natural logarithm (Parr et al. 2022, 19). This is in contrast to the unit bits, which is based on the logarithm with base 2. Crucially, *D_{KL} is always greater than or equal to zero*, and so free energy is always *greater than or equal to surprisal*.

We now have the essential tools to analyze variational free energy. The objective is minimizing complexity and maximizing accuracy:

$$F[Q,y] = D_{KL}[Q(x) \parallel P(x)] - E_{Q(x)}[\ln P(y|x)] \tag{2.2}$$

Here F is a functional, that is, a function of a function. Complexity is represented as the D_{KL} between the approximate posterior and the prior. $D_{KL}[Q(x) \parallel P(x)]$ quantifies the informational cost of departing from prior beliefs encoded in the generative model. It penalizes deviations of the approximate posterior from the prior unless they are warranted by the data. A lower D_{KL} indicates that $Q(x)$ remains closer to the prior, preserving simplicity in the model. In fact, complexity refers to "the degree to which we must depart from our beliefs about the world in order to explain data" (Parr et al. 2022, 268). This is balanced by accuracy represented by the expected log likelihood of the data y given x, $E_{Q(x)}[\ln P(y|x)]$. This term indicates how well $P(y|x)$, the likelihood of observing y given hidden states x, is supported by the approximate posterior $Q(x)$. Higher accuracy improves the alignment between $Q(x)$ and the observed data y. The negative sign reflects the mathematical structure of the functional, balancing accuracy and complexity. F is reduced by increasing accuracy without penalizing complexity.

The first term, $D_{KL}[Q(x) \,||\, P(x)]$, is always nonnegative and penalizes deviations of $Q(x)$ from the prior $P(x)$, thereby increasing F. Assuming discrete variables, the second term, $E_{Q(x)}[\ln P(y|x)]$, is an expectation over logarithms of probabilities. Since probabilities lie in the interval between 0 and 1, their logarithms are non-positive, and the expectation is therefore non-positive, becoming zero only in the degenerate case where the data are predicted with certainty. The negative sign reflects the convention of expressing free energy as a quantity to be minimized. When the expected log-likelihood is low—that is, when the model assigns low probability to the observed data under the states favored by $Q(x)$—$E_{Q(x)}[\ln P(y|x)]$ becomes more negative. As a result, its negation increases the value of F, indicating poor predictive accuracy. In this way, free energy increases when the model fails to account for the data, balancing predictive accuracy against the complexity penalty imposed by the D_{KL}.

Moreover, another important point is that $-E_{Q(x)}[\ln P(y|x)]$ is computationally tractable, assuming the likelihood $P(y|x)$ and the approximate posterior $Q(x)$ are well defined and manageable. For instance, if $Q(x)$ is a Gaussian and $P(y|x)$ is Gaussian as well, the expectation can often be computed analytically. However, this alone does not minimize free energy. The computational tractability of equation 2.2 simply allows us to calculate a specific value of F. To actually minimize F, we need to implement an algorithm—such as gradient descent—to iteratively adjust the relevant variables. The equation itself merely represents the conceptual structure of F, not the process of minimizing it—it allows us to understand what F is actually quantifying. (See https://pymdp-rtd.readthedocs.io/en/latest

/notebooks/free_energy_calculation.html to compute the variational free energy for discrete generative models, composed of categorical distributions; see also Smith et al. 2022.)

Thus, equation 2.2 implies that beliefs encoded by the approximate posterior must be adjusted to improve the accuracy with which the model predicts the observed data, while avoiding unnecessary departures from prior expectations. Minimizing free energy therefore implements a trade-off between predictive accuracy and complexity. Although variational free energy provides an upper bound on negative log evidence, the evidence itself is not computed directly; instead, minimizing free energy tightens this bound by reducing the divergence between the approximate and true posterior.

For example, you observe a temperature of 29°C and aim to infer the weather. The possible weather conditions—sunny, cloudy, and rainy—are hidden states you cannot directly observe. The likelihood $P(y|x)$ specifies how likely it is to observe 29°C for each weather condition: it is highly likely if the weather is sunny, moderately likely if it is cloudy, and extremely unlikely if it is rainy. Your beliefs about the weather, represented by the posterior distribution $Q(x)$, estimate a 60 percent chance of sunny weather, a 30 percent chance of cloudy weather, and a 10 percent chance of rainy weather. The accuracy term, $E_{Q(x)}[\ln P(y|x)]$, quantifies how well your beliefs align with the observed data. It does so by combining two factors: the likelihood $P(y|x)$, which measures how well each possible weather condition predicts the observed temperature, and your belief $Q(x)$, which reflects your confidence in each weather condition. Mathematically, it calculates a weighted average of the log-likelihood values across all possible weather conditions, where the weights are determined by $Q(x)$. This ensures that conditions you consider more plausible (e.g., sunny) contribute more to the accuracy calculation. For instance, since $P(y|x)$ is high for sunny weather and $Q(x)$ assigns significant probability to sunny weather, the contribution of sunny to the accuracy is substantial. In contrast, rainy weather has both a very low likelihood and a low belief weight, contributing minimally. By combining these factors, the accuracy term reflects how well the model as a whole explains the observed data $y = 29°C$. A higher accuracy indicates that the model aligns well with both the data and your beliefs, making it a key measure of the model's explanatory consistency.

Now, F can also be represented as divergence versus surprise (Friston 2009, 294), given by the following formula:

$$F[Q,y] = D_{KL}[Q(x) \parallel P(x|y)] - \ln P(y) \tag{2.3}$$

In this case, the D_{KL} refers to the divergence between the approximate posterior Q and the true posterior $P(x|y)$. Essentially, a low value of F indicates that we are successfully adjusting our approximate posterior to better account for the data under the assumed generative model. As the divergence decreases, "the free energy approaches the negative log evidence (surprise)—and becomes equal to surprise, if the approximate posterior Q matches the exact posterior $P(x|y)$" (Parr et al. 2022, 30). Minimizing F means reducing the divergence between the recognition density and the true posterior, allowing the two to converge and, ultimately, become equivalent. In such cases, F simplifies to surprisal, making surprisal minimization with an accurate and fully tractable model a limiting case of free energy minimization. However, in most scenarios, the free energy minimizer operates within constraints—it can only approximate the true posterior as closely as possible within the limits of a restricted, computationally tractable class of simpler distributions.

Let us take stock. The fundamental problem is that computing $P(y)$ is not computationally tractable. The computation of $P(y)$ involves integrating or summing over all possible values of x, which is often high-dimensional and intractable. If x is continuous, the integral is rarely possible to compute exactly, except in cases where conjugacy properties allow for an analytical solution (e.g., in Gaussian models with conjugate priors). In most cases, numerical integration is infeasible due to the high dimensionality of the latent variable space. A standard way to work around this is variational inference: we introduce an approximate posterior $Q(x)$ and optimize a tractable functional—the variational free energy—defined as $-E_{Q(x)}[\ln P(y|x) + \ln P(x) - \ln Q(x)]$ which is an upper bound on $-\ln P(y)$, becoming equal to $-\ln P(y)$ only when $Q(x)=P(x|y)$.

Equation 2.3 shows that, by minimizing free energy, this bound is tightened, and in the limiting case where the approximate posterior matches the true posterior, free energy becomes equal to surprisal. Although the true posterior $P(x|y)$ cannot be computed efficiently—since it depends on the intractable marginal likelihood $P(y)$—it can be approximated using a variational distribution $Q(x)$. Since $-\ln P(y)$ does not depend on Q, minimizing free energy therefore corresponds to minimizing the $D_{KL}\left[Q(x) \| P(x|y)\right]$ within the constraints of a tractable variational family.

Crucially, equations 2.2 and 2.3 are not distinct methods for minimizing F but are representations of distinct aspects of F: equation 2.2 represents that minimizing F requires finding a certain kind of trade-off, while equation 2.3 makes explicit that F is an upper bound on surprisal. We cannot find the optimal trade-off in 2.2 without automatically minimizing the D_{KL} in 2.3, and vice versa. It is crucial to recognize that accuracy, surprise, and evidence are distinct but related concepts in active inference. Accuracy refers to how well the model explains the observed data through its likelihood function $P(y|x)$. Surprisal (negative log-evidence) measures how unexpected an observation is under the model. Marginal likelihood (evidence) is computationally intractable, which is why active inference optimizes F as a tractable bound. This distinction is significant because it affects how the two equations (2.2 and 2.3) should be interpreted. Both exhibit a fundamental tension structure between accuracy and prior complexity, as well as between divergence and evidence. The goal is to balance data fidelity (ensuring the model captures relevant information) with consistency with prior expectations. It is also important to note that in equation 2.2 increasing accuracy does not necessarily minimize D_{KL}. A model may fit the data extremely well but deviate significantly from its prior, increasing complexity. Likewise, reducing D_{KL} ensures adherence to the prior but does not necessarily enhance explanatory power—if a posterior remains too close to the prior, it may fail to capture the data properly. Thus, these two measures must be carefully balanced to effectively minimize free energy. Equation 2.3 presents a slightly different scenario. F serves as an upper bound on surprisal—equation 2.2 also does this, but it does so by formulating the problem in a different way. In equation 2.3, minimizing D_{KL} minimizes free energy, improving inference even without direct access to the true posterior or surprisal. When the approximate posterior matches the true posterior, F reduces to surprisal, making surprisal minimization a special case of free energy minimization.

For a more technical discussion, see Smith et al. (2022), which focuses on the mathematical structure and the construction of partially observable Markov decision processes (POMDPs) for empirical data analysis, and Bogacz (2017), which explores the neural implementation of active inference.

Now, considering F as the measure of inaccuracy to be minimized, we can summarize our discussion as follows (Mann et al. 2022):

$$\text{Variational Free Energy Principle: } Q(x) \leftarrow \underset{q}{\operatorname{argmin}} F \tag{2.4}$$

In this formula, $\underset{q}{\operatorname{argmin}}$ means choosing the variational distribution that minimizes F, thereby maximizing model evidence (i.e., self-evidencing).

What strongly characterizes active inference is the fact that the inferential process is not limited to identifying the most probable hidden causes of observations, but is also about actions and their consequences, that is, choosing how to explore and transform the world. In other words, active inference and the FEP serve as a model for decision-making. To better understand this concept, we first need to introduce the notion of expected free energy (EFE, or G). In this case, the problem is reversed: it is no longer about identifying the most probable hidden causes of observations, but about identifying the most probable future consequences (i.e., sensory data, observations) of actions—the cause is already known, being the system itself. The question is: How does the system select the sequence of actions (i.e., future states) that most closely matches its preferences, that is, its priors? If we put the dog in the bedroom, where do we expect it to be? In this case, the situation is reversed: we are the cause of the observations. Therefore, the most probable behavior of an active inference agent is the sequence of actions and observations that minimizes G more effectively than any other. We can model the behavior of an active inference agent using only two key variables: VFE and EFE. Probabilistically, an active inference agent tends to select perceptions and actions that, overall, minimize both of these values.

In conclusion, we can encapsulate these ideas with the following quote from Ramstead et al. (2021, 234):

> Active inference generalizes approximate Bayesian inference, since in active inference the objective is not simply to infer the hidden states that cause observations but, more importantly, to act in such a way that minimizes self-information or surprise (via minimizing free energy) or minimizes expected surprise or uncertainty (by minimizing expected free energy). The reason for this is simple: active inference turns on the idea that it is action, upon which perception depends, that ultimately minimizes uncertainty about the external causes of sensory observations. Hence, action can be cast as placing an upper bound on surprise—and expected surprise or uncertainty.

From this, we clearly see how active inference and the FEP provide a novel interpretation of one of the most important ideas of cybernetics, that is, the good regulator theorem, originally proposed by Conant and Ashby (1970). It elucidates the relationship between a control system—embodied by the generative model with its priors—and the system being controlled, namely, the organism and its adaptive behaviors within the world. Active inference emerges from a pragmatist reinterpretation of the theorem. Indeed, the good regulator theorem posits that a system can effectively control another only if it is isomorphic to the critical properties of the system it governs—that is, if it functions as a statistical model of the system's relevant features (Conant & Ashby 1970). Generative models accomplish this by mirroring the structure of the generative processes they regulate, thereby enabling the control of an organism's behavior. Rather than aiming to represent an external reality beyond sensory impressions, generative models prioritize the regulation and guidance of action (Bruineberg et al. 2016). Their primary goal is not the pursuit of objective truth but the facilitation of survival. By modeling the organism itself, these generative models enable systems to adaptively modulate their behavior in response to dynamic environments. The generative model serves a pivotal role in guiding actions that are finely attuned to context. This highlights the need to shift our focus from merely understanding how brain mechanisms, such as Bayesian predictive coding, build and sustain models of the world to exploring how these mechanisms drive feedback loops that foster environmental attunement and support adaptive behavior (Ramstead et al. 2021, 234). Environmental attunement, in this context, means that organisms survive not by constructing a detached, objective map of reality but by dynamically adjusting their generative models to regulate behavior in ways that maximize adaptability within their ecological niche (figure 2.4).

2.8 HIERARCHICAL MODELS

As mentioned, active inference scholars differentiate between two critical types of free energy: variational free energy and expected free energy. These concepts lie at the heart of the framework. Now, let us take a closer look at how they relate to one another.

A

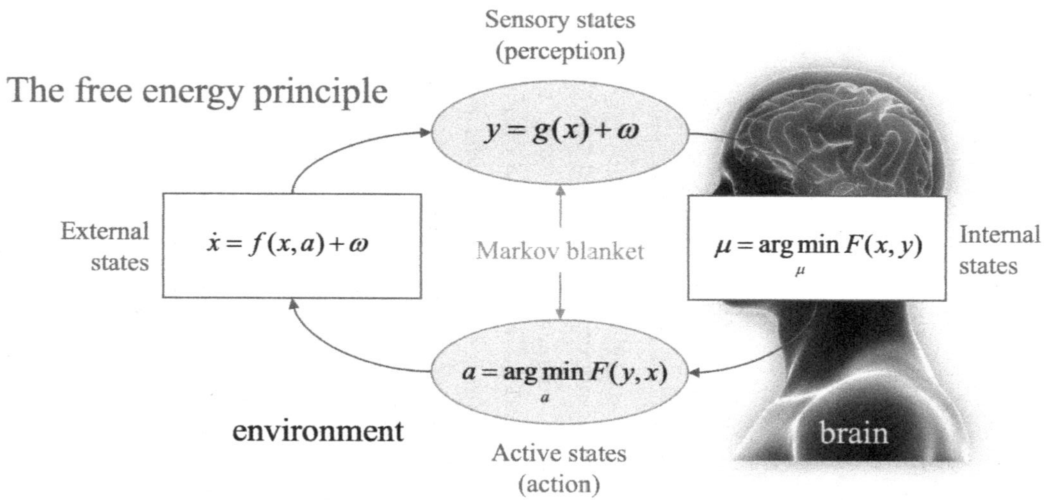

$$F = Energy - Entropy = -\langle \ln P(x, y) \rangle_Q + \langle \ln Q(x) \rangle_Q$$

B

Action to minimize a bound on surprise

$$F = Complexity - Accuracy$$
$$= D(Q(x:\mu) \,||\, P(x)) - \langle \ln P(y(a) \mid x) \rangle_Q$$
$$a = \arg\max_{a} Accuracy$$

←

Perception to optimize the bound

$$F = Divergence + Surprise$$
$$= D(Q(x:\mu) \,||\, P(x \mid y)) - \ln P(y)$$
$$\mu = \arg\min_{\mu} Divergence$$

Figure 2.4

(A) A schematic representation (from Ramstead et al. 2018) of the core components of the free energy principle, illustrating the relationship between a system's internal states (μ, e.g., the brain), its sensory and active states (the Markov blanket), and its external environment. The system receives sensory inputs ($y = g(x) + \omega$) generated by hidden external states (x), which evolve according to a dynamical process ($\dot{x} = f(x, a) + \omega$), where ω represents random fluctuations. Actions (a) emitted by the system affect the environment and its future sensory inputs. Internal states and actions work jointly to minimize variational free energy—a bound on sensory surprise—by forming and updating a probabilistic belief distribution $Q(x)$ over hidden states. The Markov blanket mediates this interaction by statistically shielding internal states from direct influence by external states, enforcing conditional independence. Sensory states depend only on external causes, while active states depend only on internal dynamics. (B) Two complementary aspects of free energy minimization. Perception updates internal states to reduce divergence between the approximate posterior $Q(x : \mu)$ and the true posterior $P(x|y)$, refining the internal representation of hidden causes. Action, by contrast, optimizes sensory sampling: it selects behaviors that increase expected accuracy by fulfilling predictions, effectively reducing the discrepancy between expected and actual sensory input. Together, these processes minimize free energy as a trade-off between complexity and accuracy, allowing the system to sustain a low-surprise existence in a changing environment. *Source:* M. Ramstead, P. Badcock, and K. Friston, 2018, "Answering Schrödinger's Question: A Free Energy Formulation," *Physics of Life Reviews* 24, pp. 1–16.

First, a short outline:

- Variational free energy (VFE) is fundamentally about the fit between a model and sensory data, operating on current and past observations. It is minimized by optimizing the alignment between predictions and actual sensory inputs with respect to internal (belief) states. This minimization process allows the generative model to increasingly fine-tune the recognition density to align with the generative process.
- In contrast, expected free energy (EFE) is utilized in planning and decision-making, focusing on prospective outcomes and future states. Unlike variational free energy, expected free energy involves a consideration of future, policy-dependent observations, evaluating how policies might influence states and observations over time. This makes EFE a forward-looking measure that helps an organism plan by anticipating the outcomes of different actions and choosing the ones that are expected to minimize future surprise.

This distinction between two types or levels of active inference is crucial for several reasons. First, it deepens the concept of active inference by introducing a temporal dimension. Second, it enhances clarity by providing a more structured articulation of hierarchy. Expected free energy always precedes, guides, and regulates variational free energy. We cannot discuss the latter without considering the former. This difference in levels also highlights another crucial point: the process of precision evaluation is always the outcome of an inferential process and cannot exist independently of it. As mentioned in chapter 1, the precision-weighting mechanism is a crucial element: the active inference agent can make predictions and update beliefs not only about sensory inputs but also about their variability, and thus the precision of both predictions and prediction errors.

Let us now focus on the EFE. Active inference is not just about minimizing prediction errors in perception; it also serves as a framework for action selection. When determining optimal actions, it is not enough to consider current observations alone; actions must be selected to achieve desired future outcomes. EFE deals with policies (π), that is, sequences of possible actions in a given context. A policy is a hypothesis about a course of action. Associated with each policy is "a trajectory of hidden states" (Parr et al. 2022, 32).

Now, to properly understand the distinction between VFE and EFE, one must understand the change in crucial elements that occurs in the transition between the two. Whereas in VFE the system (e.g., an organism) has the data and then makes hypotheses about its possible unobservable causes from that data, in EFE the organism doesn't know the data. This means that the organism has to imagine the data, the observations, because they have not happened yet—they lie in the future. Therefore, the organism develops a policy and tests it, that is, predicts possible sensory outcomes of that policy (i.e., *what if?* What are the possible consequences of this sequence of actions?) Each policy is thus given a rate, and based on this the organism calculates the posterior (i.e., the best prior, which is the motive for the action and thus the cause of the expected observations). Thus, EFE is essentially probabilistic counterfactual reasoning aimed at selecting the best policy in a given situation. In other terms, this means that "to infer optimal actions, a model must predict sequences of future states and observations for each possible policy, and then calculate the EFE associated with those different sequences of future states and observations" (Smith et al. 2022, 10).

The following is one of the classic formulations of G (from Parr et al. 2022, 33, slightly modified):

$$G[\pi] = -E_{Q(x,\,y\,|\,\pi)}[D_{KL}[Q(x\,|\,y,\pi)\,||\,Q(x\,|\,\pi)]] - E_{Q(y\,|\,\pi)}[\ln P(y\,|\,C)] \quad (2.5)$$

G evaluates the utility of a policy π by integrating two key components: information gain and pragmatic value. The first term measures the expected D_{KL} between the prior $Q(x|\pi)$ (before making an observation) and posterior beliefs $Q(x|y,\pi)$ under the policy π (after making an observation). This can be reframed as the question: How much do I expect my beliefs to change (or update) after making a new observation under this policy? The negative sign indicates that the greater the update to my beliefs, the more effectively G is minimized, thereby maximizing my information gain—I want observations that provide information, that is, observations that reduce uncertainty and distinguish among relevant hypotheses. The second term evaluates how well the outcomes y align with the agent's preferences, encoded as a prior distribution over outcomes $P(y|C)$ (this is a crucial move; see Smith et al. 2022, 10). As said, the negative signs ensure that minimizing G corresponds to maximizing both information gain and the alignment of outcomes with preferences. This dual optimization leads to policies that balance exploration (reducing uncertainty) and exploitation (achieving desired outcomes); see Parr et al. (2022, 33–34) and Millidge et al. (2021, 18–19). We may also integrate additional factors, such as habits or expected precision of G, that is, the confidence in the policy. When confidence is lower, habits exert a stronger influence (Smith et al. 2022). Expected precision values can be updated based on changes in confidence after new observations.

The goal of the decision process should be to choose the π that minimizes G more, such that interventions and choices bring the posterior beliefs about future states of the world closer to prior preferences. In other words, the policy that minimizes G the most is the one an active inference agent is most likely to choose. The process is reversed in comparison to VFE: we do not reason about hidden causes of observations, but about future observations (not visible, not present) produced by known causes (actions of the system). The priors play a special role. These are the priors over states in the long-term future and are distinct from the priors over hidden states during inference. In other words, the priors in the VFE are distinct from the priors in the EFE. The key thing is that VFE is a functional of approximate posterior beliefs and is used to optimize those beliefs. This is completely different from the use of EFE, which is a function of policies or interventions and is therefore used to identify the most likely choice, decision or intervention. Again, we use negative log evidence to measure how much the future data will surprise the agent on average. More precisely, in this case we do not consider the negative log evidence, but rather the expectation of the negative log evidence—the expected surprise of future observations given the occurrence of a state. G scores different policies according to their ability to reduce uncertainty or expected surprise (i.e., fulfill prior preferences while resolving ambiguity).

Considering G as the measure of inaccuracy to be minimized, the solution is as follows:

$$\text{Expected Free Energy Principle: } \pi \leftarrow \underset{\pi}{\operatorname{argmin}}\, G \tag{2.6}$$

Now suppose that the organism processes its counterfactual reasoning and comes to choose the most likely policy, that is, the sequence of actions that reduces surprise the most relative to its preferences. The counterfactual becomes a motive for action. The action generates certain observations. These observations are evaluated by the organism itself through VFE, that is, as sensory data whose cause must be found. In other words, the organism acts and asks itself whether the changes it observes in its environment are those predicted by its policy or not. To put it more simply, the prediction with a wide time range (policy) must be confirmed step by step by predictions with a narrower time range (actions). The key point is the relationship between the two types of predictions: policy versus actions. If the organism calculates that the EFE of a policy is very low, it means that it has great confidence in this policy, that it considers it reliable, accurate. If this is confirmed by actions and perceptions, then this policy has a high level of reliability. If, on the other hand, it is disproved, in the sense that the VFE connected to that policy (i.e., better: associated with the posterior beliefs used to track the policy step by step) tends to increase, then there is an error, and the inferred value of that policy must be revised. The model is hierarchical: more abstract levels of probabilistic planning (EFE) simultaneously influence and depend on more concrete, present-bound levels of probabilistic reasoning (VFE). The more important the policy (in the sense that it has a deep, extended temporal scope), the more relevant the error is and the more it affects multiple levels. Thus, it is important to note that precision estimates reflect inferred confidence in policy selection, modulating the influence of EFE on action selection. Therefore, precisions emerge from a process in which: (1) prediction errors (VFE) update confidence in beliefs; (2) expected policy values (EFE) are weighted by precisions; and (3) precision itself has a hyperprior and its own dynamics. The value of a policy is constantly changing according to the relationship between these hierarchical levels.

Another fundamental point that should be emphasized is the distinction between different types of generative models (Parr et al. 2022, chs. 7 and 8). There are two primary types of generative models that are particularly prominent in active inference: those using discrete time formulations, like partially observable Markov decision processes (POMDPs), and those using continuous time formulations, often represented through stochastic differential equations. These models are adapted based on whether the environmental variables or system states they are modeling are better understood as evolving in discrete steps or as continuous processes. Discrete time models, such as POMDPs, are highly relevant when addressing problems where decisions or changes occur in response to discrete, observable changes in an environment. This makes them suitable for scenarios like navigating through a maze or making sequential choices, where each decision point is distinct and impacts future possibilities. In contrast, continuous time models are essential when dealing with variables that change in a fluid, ongoing manner. These models often utilize differential equations to describe phenomena like motion or changes in belief across time, providing a smooth trajectory that predicts future states from current and past conditions. Each type of model brings its own strengths to different cognitive and computational challenges, emphasizing the need to select the model type based on the nature of the problem being tackled (Maisto et al. 2025; Smith et al. 2022; Wei 2024).

A concrete example may be helpful. Harris et al. (2022) applied active inference to sports, modeling how athletes predict and respond to dynamic and stressful situations, such as a fastball in baseball or a penalty kick in soccer. Using computational models grounded in Bayesian principles, the researchers examined how athletes integrate contextual cues and real-time sensory inputs to optimize decision-making under uncertainty. Their simulations utilized POMDPs to represent scenarios where athletes must act on incomplete and evolving information. The study simulated two key scenarios: one where athletes received early sensory cues and another where cues arrived later, forcing a trade-off between acting quickly with limited data or waiting for more precise information at the cost of reaction time. The findings revealed that athletes rely heavily on prior contextual knowledge, particularly when sensory data is weak or ambiguous. As sensory inputs become clearer and more reliable, they increasingly dominate decision-making. These results align with empirical research in sports psychology, highlighting the adaptive integration of context and sensory data. The simulations demonstrated key trade-offs: acting early risks inaccuracy, while waiting risks setbacks. Over time, athletes refine their anticipatory skills, prioritizing reliable information and strategically positioning themselves to reduce uncertainty and simplify decisions. This study showcases how active inference models can effectively capture the nuanced dynamics of anticipatory behavior in fast-paced environments.

2.9 A NEW *DESIGN FOR A BRAIN*

This analysis gives me the opportunity to better clarify the concept of third cybernetics that I evoked at the beginning of this book. It is noteworthy how Dupuy (2009), in his philosophical and historical reconstruction of cybernetics, aligns the crisis of early cybernetics with Ashby's "return to metaphysics." In this view, Ashby's work represents a kind of turning point, beyond which the cybernetic project could no longer remain the same—and that was the beginning of the end of the movement. Regardless of Dupuy's (2009) philosophical interpretation, it is important to notice that Ashby is a key reference for scholars of PP and active inference. It is no coincidence that a seminal paper like "Whatever Next? Predictive Brains, Situated Agents, and the Future of Cognitive Science" (Clark 2013) opens with a quote from Ashby: "The whole function of the brain is summed up in: error correction." Similarly, Seth (2015, 7) explicitly lists Ashby among the precursors of the FEP: "The close coupling of perception and action emphasized in the free energy principle points to a deep connection between predictive processing and mid-twentieth-century cybernetics." Seth (2015) highlights a central idea from Ashby that directly influences both PP and the FEP: adaptive systems are those capable of maintaining a set of essential variables within certain limits across a wide range of environments. This idea lays at the heart of the Homeostat, a machine Ashby presented during the Ninth Macy Conference (March 20–21, 1952), which provoked considerable criticism and skepticism from other cyberneticists. The Homeostat was an unusual machine—it had no practical purpose and performed no specific tasks. Instead, it embodied some core hypotheses from Ashby's book *Design for a Brain* (1952) that can be summarized in two aphorisms. The first: "The brain organizes nothing—it acts" (n. 50), and the second: "The brain is nature's latest and ferocious instrument of self-preservation" (n. 28).[2] These are ideas that any proponent of active inference would readily embrace.

Do not forget that Ashby was a neuropsychiatrist focused on understanding the nature of mental illness and pathological behavior. This goal underpins the purpose of *Design for a Brain*: developing the concept of the abstract machine, a purely mathematical construct, as the key to understanding the brain, as well as any system exhibiting self-organizing and adaptive capacities. In Ashby (1952), the central question is: How can a machine induce self-organized changes in its structure? The answer involved two distinct sets of variables:

- Machine 1: A set of variables that change continuously over time.
- Machine 2: A set of variables that remain constant for long periods but shift rapidly and discretely between values when needed.

These two sets are coupled in such a way that as long as Machine 1's variables remain within a stable range, the variables of Machine 2 do not change. However, when Machine 1's variables move outside this range, Machine 2's variables switch to a new value, creating a different pattern of interconnections among the variables and thus a new organization. If the new organization fails to stabilize Machine 1's variables within a certain range of values, Machine 2 will shift again, producing another reorganization. This cycle repeats until stability is achieved: Machine 2 continues to adjust until Machine 1 reaches equilibrium, that is, a steady state of values. Stability is thus defined as an organization C in which, once the system enters C, it will remain there until the next perturbation. The system actively resists disturbances that threaten to move it outside C by initiating counteractive adjustments (triggered by changes in Machine 2's variables).

On this basis, Ashby distinguishes between a stable system and an ultrastable system:

- A stable system consists solely of continuous variables (Machine 1). It remains within its stable field as long as no disturbance pushes it outside. However, if pushed out, it cannot return to stability on its own.
- An ultrastable system contains both sets of variables (Machine 1 and Machine 2). This allows it to reorganize itself, adapt to perturbations, and recover or redefine a stable configuration. The Homeostat is an ultrastable machine. There is no single, absolute stable configuration—it always is dynamic and context-dependent.

Two key aspects emerge clearly from these observations. First, Ashby introduces randomness into the machine. The discrete variables in Machine 2 are assigned values entirely at random, in a non-deterministic, unprogrammed way. In other words, Machine 2 operates through trial and error, attempting to restore equilibrium in Machine 1 without following a predetermined plan. Second, the interaction with the environment is essential. Comparing the Homeostat to the Turing machine reveals two fundamentally different ways of conceptualizing machines—and, by extension, the brain. Turing's approach is *instructionist*: the machine is defined by its instructions, or program, and the environment plays a negligible role in its operations (Turing 1936). In contrast, the Homeostat is *selectionist*: the machine can adapt to different situations in its search for equilibrium; it independently finds adaptive solutions, meaning that the environment is integral to the machine, playing a crucial role in shaping its operations. This distinction mirrors the divide between "good old-fashioned artificial intelligence" (GOFAI)—a symbolic and representational approach to simulating intelligent behavior—and the more recent developments in AI, such as machine learning and deep learning systems (Haugeland 1985). Unlike GOFAI, the Homeostat involves no symbols or representations. Instead, it relies on patterns of action aimed at achieving a context-relative equilibrium. For Ashby, the program—a structured and stable series of operations and rules guiding action toward a goal—is not the starting point but the outcome.

Beyond the principle of ultrastability as a crucial condition for adaptive behavior—explicitly referenced in Ororbia and Friston (2024, 10–11)—another key aspect of Ashby's cybernetics lies at the heart of the FEP. It is the *law of requisite variety*. In my view, the FEP (at least in its classical formulation) can be understood as an extension of this law.

Ashby's law of requisite variety states that in order to control a system effectively, the variety of the regulator must be equal to or greater than the variety of the disturbances affecting the system. Variety refers to the number of possible states a system can take. Control is achieved by matching the regulator's capacity to produce sufficient responses to the range of disturbances, ensuring that the system remains within desired parameters. The law highlights the fundamental requirement for balance in complexity between a controller and the system it regulates to ensure stability and adaptation in dynamic environments (Ashby 1956). This law also echoes one of the key aspects of the FEP. As mentioned, free energy is an upper bound on the negative log evidence, or surprisal, expressed as $F \geq S$. Unlike Ashby's original formulation, which primarily emphasized reactive control, the FEP introduces an anticipatory dimension— but this is just one way to reinterpret the principle. Adaptive systems governed by the FEP do not merely react to disturbances; they predict them. The generative model must be more flexible than the environment (i.e., more able to update quickly and thus take on more configurations) in order to control the environment and thus minimize uncertainty. This also means that "for a system to exert sufficient control over its environment, it should try to maximize its internal variety to be optimally prepared for perturbations that it might encounter; maximizing internal entropy increases its cybernetic freedom or options for action" (Ororbia & Friston 2024, 11)—this is the core of EFE.

Thus, the FEP extends Ashby's principle beyond static balance, emphasizing the proactive minimization of uncertainty through generative models. Furthermore, the FEP introduces a hierarchical and multiscale perspective.

Ashby was a pivotal figure for both first and second cybernetics—almost a bridge between them. The connections between his work, active inference, and the FEP not only provide a valuable historical perspective but also offer a philosophical lens to better understand and interpret these contemporary frameworks. Can we speak of third cybernetics—a new version of the original cybernetic project, envisioned as a physicalist science of the mind? Could active inference and the FEP—"a physics-based approach to intelligence that aims to describe, study, and design intelligent agents *from their own perspective*" (Friston et al. 2024, 6)—represent a grand interdisciplinary endeavor aimed at restoring and revitalizing the key concepts of first and second cybernetics from a fresh perspective? Perhaps it is still too early to say. As Dupuy (2009) compellingly argued, cybernetics was a failure born of brilliant ideas that led to the birth of cognitive sciences (although this origin was later erased, denied, or repressed). A return to the authentic spirit of cybernetics is now possible, despite this collective repression. There is no denying that from Friston's (2005) seminal work, "A Theory of Cortical Responses," to Ororbia and Friston (2024), active inference has been progressively expanded into a comprehensive framework for, as Friston et al. (2024) titled it, "Designing Ecosystems of Intelligence from First Principles"—a title that resonates unequivocally with Ashby's *Design for a Brain*.

2.10 SOME CRITICISM, AND A NEW IMAGE FOR THE FEP

The debate surrounding active inference and the FEP is vast, encompassing enough material for a book of its own. Here, I concentrate on a select few critiques that are most pertinent to the aims of this chapter. It is important to note that purely philosophical perspectives on the FEP—whether supportive or critical—risk neglecting its critical foundation in empirical validation and its origins in actual experimentation and testing, rather than mere theoretical abstraction (Isomura et al. 2023).

One of the most common criticisms of the FEP is its triviality—that is, its extreme generality allows it to be applied to anything, and thus it offers no real epistemic advantage: explaining everything ends up explaining nothing (Colombo & Wright 2018; van Es 2020; Gershman 2019; Gershman

& Daw 2012; Nave 2025; Sims 2016). These critiques argue that the FEP lacks biophysical or cognitive realism, contradicts experimental observations (it is based only on synthetic data), cannot adequately explain brain function, or fails to generate predictions. But the "real worry" is that "the FEP is somehow empty; that it lacks all conceptual content" and is unfalsifiable (Andrews 2021, 30). Actually, a certain tendency in the FEP literature toward reification and universalization—treating the FEP as an ontological structure of reality universally applicable—reinforces this position.

This critique should not be underestimated, though three points must be kept in mind:

1. Active inference and the FEP are still relatively young and evolving frameworks, born out of the remarkable advances in neuroscience over the past three decades: "The FEP is a living theory, subject to internal controversy, change and revision" (Millidge et al. 2021, 1). Although they hold great promise and can count on tons of empirical data (Friston 2022 discusses this point), these theories rest on a mathematical foundation that is still developing. The absence of a standardized formalism has led to the emergence of multiple, sometimes competing, interpretations of the core principles, each emphasizing slightly different aspects. While this diversity reflects the flexibility and richness of these frameworks, it can also render them inaccessible or overly complex for those unfamiliar with the field. For researchers and practitioners outside the immediate domain, this variability can come across as esoteric or opaque, potentially hindering wider acceptance and application.
2. To address the accusation of triviality, it is essential to clarify the epistemic nature of the FEP. As mentioned earlier, considering the FEP as an empirical theory is a category mistake. Drawing on Andrews (2021) and Weisberg (2013), I propose to distinguish between *model structure* and *model construal*. Scientific models consist of an abstract structure—representing objects, states, relationships, and constraints—and an interpretation that assigns meaning to this structure in specific contexts and disciplines. Weisberg identifies three types of model structures: concrete, mathematical, and computational. Mathematical structures are expressed through descriptions such as equations, algorithms, or diagrams, which complement the structure but do not constitute the model itself. Interpretation, or construal, gives the structure semantics by relating it to

real-world phenomena and evaluating its explanatory power through assignment, scope, and fidelity criteria. Applied to the FEP, this distinction reveals that the principle itself is only a mathematical structure. Consequently, as Andrews (2021) claims, the FEP is merely a mathematical framework that provides a foundation for further applications. It has no inherent semantics, and, therefore, it cannot provide any meaningful information about anything. FEP scholars might counter by arguing that its triviality precisely underscores its self-evident nature as a foundational principle.

3. Andrews' (2021) position can be reinforced by the reflections of Millidge et al. (2021, 20), who argue that the FEP is a mathematical structure that can model dynamical NESS systems. Crucially, the FEP in its most general formulation "does not make any specific predictions about the dynamics of the system" (Millidge et al. 2021, 20). Although "several process theories have been explicitly derived from within the FEP framework, all such theories necessitate making specific and ultimately arbitrary modelling choices, such as of the generative model and variational density" (Millidge et al. 2021, 20), these choices "sit below the level of abstraction that the mathematical theory of the FEP exists at" (Millidge et al. 2021, 20). However, according to Millidge et al. (2021, 21), the real issue with the FEP lies not so much in its falsifiability but in its *applicability*—that is, in its premises, the conditions a system must satisfy for the FEP to apply (having a NESS and Markov blanket). It is unclear if "real systems of the kind the FEP desires to explain—such as biological self-organization, and especially brains, can fulfil [these conditions]" (Millidge et al. 2021, 20). If it turns out that such systems violate the conditions, "then the FEP cannot be said to apply to them and thus cannot be of use in understanding them, even as an interpretatory device" (Millidge et al. 2021, 20). In other words, while the FEP—as a purely mathematical formalization of certain dynamical systems—may not be falsifiable, its applicability could be highly limited, confined to a specific class of systems that exhibit those properties. Importantly, "many of the assumptions of the FEP, when interpreted strictly, do not appear to hold in general for complex biological systems such as brains" (Millidge et al. 2021, 20). This is the real problem that needs to be solved.

Nave (2025) provides a useful account of the state of the literature between 2012 and 2023, and a useful high-level overview of the NESS formulation of the FEP. But the book overgeneralizes from that (admittedly limited) classical formulation and fails to consider the deep relationships between information theory and thermodynamics that the FEP formalizes. Nave (2025) acknowledges that the FEP offers a compelling account of predictive control and system stability, yet she argues that it is trivial and unable to distinguish living from nonliving systems. Drawing on cybernetics, PP, and phenomenology, Nave emphasizes that biological organisms depend on continuous material turnover and metabolic activity to maintain their structure, whereas engineered systems like the UK National Grid's Electricity Control Centre (ECC) merely resist change and resume operation when external inputs return. She introduces the notion of "constraint closure" to capture the reciprocal dependencies that confer autonomy and intentionality on living cells, suggesting that without this concept, the FEP cannot account for life's intrinsic processual dynamics. In a vivid example, Nave contrasts the ECC's transistors and wiring—stable in the absence of power—with a cell's enzymes and membranes, which irreversibly disintegrate without sustained metabolism. The FEP is unable to address the precarious self-sustaining loops that define life. Nave (2025) therefore positions the FEP not as a "theory of everything," but as a useful mathematical tool whose explanatory power must be supplemented by biological concepts.

While Nave's critique effectively challenges extreme or naïve readings of the FEP, it does not invalidate the principle itself. First, the supposed binary distinction between cellular instability and artificial stability overlooks the active maintenance required by any kind of technical systems. The ECC relies on human operators, automated monitoring algorithms, regular hardware replacement, and environmental controls to preserve its complex functionality—processes that parallel, albeit on a different substrate, the homeostatic feedback loops that sustain living organisms. All physical systems, biological or artificial, are subject to entropy, material degradation, and external perturbations; true resilience emerges from ongoing corrective activity, not from inherent structural invariance. Nave's argument seems to be that the ECC, like any technical system, has only one purpose: stabilization through resistance to disruptions—a purely reactive function. However, in my opinion, this overlooks the complexity of artificial systems. Cybernetics first introduced the idea of *telos* (purpose) into machines, showing that if we ignore this purpose, which is not simply stabilization, we miss the essence of the machine itself.

Second, Nave's claim that life must exhibit unbroken metabolic activity neglects well-documented biological strategies of dormancy and hibernation. Many organisms—spores, seeds, extremophiles—enter low-activity states in which metabolic processes are almost entirely suspended yet viability persists. Synthetic biology further blurs this distinction by engineering cells and materials capable of self-repair and dynamic feedback, challenging any rigid divide between living and nonliving systems. If the FEP appears unable to demarcate life from nonlife, this may reflect the fundamental difficulty of defining "life" rather than a flaw in the principle. Philosophical frameworks such as Levin's TAME and Latour's ANT demonstrate that natural, social, and technological domains are deeply entangled. The FEP's mathematical generality is therefore a strength: it provides a unifying lens for understanding self-organization, emergence, and agency across diverse systems.

All the FEP criticisms we have discussed—and which Nave (2025) systematically revisits—are important because they represent a specific stage in the history of the debate on active inference and the FEP. However, as I have already pointed out, these critiques have in fact been superseded by the latest theoretical developments in the literature. As mentioned earlier, several new studies (Beck & Ramstead 2025; Friston et al. 2020; Friston et al. 2023a; Ramstead et al. 2024; Sakthivadivel 2022a, 2022b, 2022c) have essentially ushered in a new phase of the debate on the FEP and active inference.

Friston et al. (2023a) and Ramstead et al. (2023) distinguish two major formulations of the FEP: the paths-based formulation and the state-based formulation. The authors clearly emphasize that the paths-based formulation (i.e., a formulation based on probability densities on the paths or trajectories) is both the most general and the original expression of the FEP. It is grounded in the variational principle of stationary action and does not require the presupposition of a NESS. "There is no need, here, for assumptions as to steady state densities or stationary modes: this is a straightforward path of least action, where autonomous states minimize their action" (Ramstead et al. 2023, 10). By contrast, the state-based formulation (i.e., a formulation based on probability densities on the states), which does rely on the existence of a NESS, is acknowledged as narrower in scope—useful in specific domains such as biological rhythms—but not necessary for the general applicability of the FEP. Moreover, the concept of the Markov blanket is also present in the path-based formulation of the FEP. The point is that if we can distinguish between internal and external paths then we will see that the internal and active paths can be described as performing Bayesian inference. In short, under the FEP, "we can say in full generality (i.e., without making any assumptions about stationarity or steady state) that the existence of a particular partition, defined in terms of sparse coupling of paths of internal and external subsets of a particular system, licenses an interpretation of the most likely internal and active paths, given sensory paths, as instantiating an elemental form of Bayesian inference" (Ramstead et al. 2023, 9). This marks a significant advance in the debate and opens the door to a new "physics of belief," which philosophy, anthropology, psychology, and sociology will be tasked with exploring in the future. In other words, the FEP can be described as a restatement of the principle of least (or stationary) action within certain boundaries.

Is this enough to avoid the criticisms we mentioned? Can we always distinguish between external and internal states in a system? Organisms as simple as bacteria cannot be described as having a Markov blanket in Friston's sense, as they constantly undergo chemical exchange with their environment due to their porous, non-absolute boundaries. However, from the perspective of active inference theorists, Markov blankets are considered fundamental to the logic of the FEP.

A way to respond to this issue is Friston's (2022) approach, which directly addresses the problem. In summary, "it is perfectly admissible for a Markov blanket at one scale to persist and thereby be perceived, inferred or measured, while its constituent states derive from Markov blankets at a finer (e.g., molecular) scale that come and go very quickly" (Friston 2022). The classic example is a flame—it cannot be described in terms of Markov blankets, yet it is undeniably a "thing." So, does the flame have a Markov blanket? The short answer is that it is a matter of scale. A "thing" is simply a collection of states, and a Markov blanket is just one type of dynamic organization of these states. "I think"—Friston writes—"the answer depends upon the scale at which the question is posed. From the perspective of the flame—or the philosopher observing flame—the blanket states could be the states of incandescent plasmas that influence (and are influenced by) states beyond the flame, including the philosopher's sensory states" (Friston 2022,

12). Conversely, "from the perspective of the molecular constituents of the plasma there is no flame, because the coupling of concern is at the fast scale of molecular fluctuations" (Friston 2022, 12). In other words, the problem is not whether the flame is a thing or not, but rather why we perceive the flame (a collection of states) as a stable object (a Markov blanket) rather than as a chaotic ensemble of processes (constituent Markov blankets that are constantly forming and collapsing). In short, "the Markov blanket is just a way of formalizing the coupling among the states of things at a particular scale" (Friston 2022, 12). So, if I understand correctly, this is the logic: If X is identifiable as a stable thing at a given scale, then X may be modeled as a set of states, or paths, partitioned by a Markov blanket at that scale. Consequently, *from the observer's perspective*, X behaves as if it minimizes the free energy of its internal states through active inference (i.e., by performing Bayesian inference over external states). However, this depends on the scale, the nature of the observer, and the observed system.

The development of the concept of Markov blankets has been at the center of renewed thinking about FEP in recent years. Sakthivadivel (2022a) offers the first rigorous treatment of how Markov blankets emerge almost inevitably in complex, high-dimensional settings. After introducing a relaxed notion of a "weak" blanket that tolerates limited, controlled exchange (so long as key system statistics remain distinguishable), Sakthivadivel shows via adiabatic arguments and large-deviations bounds that for a broad class of random dynamical systems, the probability of failing to admit even a weak blanket vanishes as the number of dimensions grows without bound. In doing so, this author clearly demonstrates that almost every physically realistic, multiscale system possesses the sort of boundary structure the FEP relies on. Building on this foundational result, Beck and Ramstead (2025) transform the FEP's abstract partitioning into an explicit, data-driven procedure. Drawing on Jaynes' maximum-caliber formulation, they cast the discovery of objects and their governing "ontological potentials" as a problem of finding time-evolving blankets. Their variational Bayesian expectation maximization (VBEM) algorithm labels each microscopic element as internal, boundary, or external—allowing those roles themselves to change dynamically—and simultaneously infers the macroscopic laws that best predict observed trajectories. Notably, when applied to Newton's cradle, a burning fuse, the Lorenz attractor, and even a simulated cellular system, it recovers intuitive object boundaries and dynamics without any hand-tuned supervision.

Together, these works shore up both the theoretical and practical foundations of the FEP. Sakthivadivel's proof of blanket ubiquity ensures that the FEP's core assumption—that internal and external states are conditionally independent given boundary states—holds almost universally in complex systems. Beck and Ramstead then provide the missing link for empirical application, showing how to detect those very blankets and extract the emergent FEP functionals that govern macroscopic behavior. By uniting a measure-theoretic validation of sparse coupling with a generative modeling toolkit for dynamic partitioning, they significantly advance the FEP as a general theory of "things."

2.11 CONCLUSION

In this chapter, I aimed to outline in some depth not only the conceptual framework of the FEP and active inference, but also the evolution of the literature and the corresponding formulations of the FEP. For a comprehensive historical review of active inference, refer to Metzinger and Wiese (2017), Pezzulo et al. (2024a), and Pezzulo et al. (2024b). To understand the theoretical framework within which active inference is situated, consult foundational works such as Knill and Pouget (2004), Rosenfeld and Anderson (1989), and especially Friston (2019). These works explore a range of themes, including the evolution of neurocomputation and neurosimulation, the rise of the Bayesian brain hypothesis, and the notion that neural systems are intrinsically active, stochastic, and function through interactions among neuronal populations. This theoretical progression builds on Edelman's (1987) shift toward neural Darwinism, which marked a significant departure from genetic determinism. Edelman's work rejected both genetic determinism and the phrenological hypothesis—the idea of a direct, one-to-one correspondence between brain regions and cognitive functions—ushering in the concept of a plastic brain shaped by group selection and dynamic neural mapping.

3

TECHNOLOGY AS ACTIVE INFERENCE

This chapter and chapter 4 act as a bridge between the more theoretical part of the book (chapters 1 and 2) and the more applied part (chapters 5 and 6). This chapter is organized into two parts. In section 3.1, I review the literature that extends active inference and the FEP to the social sciences, and I advance a precise critique: those studies have overlooked the crucial role of technology. In section 3.2, I show how the FEP provides an elegant theoretical model for justifying the attribution of agency to artifacts. I bolster this claim by showing how key concepts from the philosophy of technology, actor-network theory (ANT), and technology-focused cognitive science can be reinterpreted—and even enhanced—through active inference and the FEP. Based on this, chapter 4 will develop a roadmap for a FEP-based design methodology.

My argument is as follows. In active inference, agency is essentially self-evidencing (Hohwy 2016)—an emergent property of any physical system endowed with a Markov blanket. The very presence of a Markov blanket (i.e., a particular partition of states that imposes conditional independence between internal and external states, given the blanket states) entails agency as self-evidencing—this is by no means trivial since a Markov blanket is not just a label you can arbitrarily assign to any system (Beck & Ramstead 2025; Hesp et al. 2021; Thestrup Waade et al. 2025). Essentially, the FEP can be seen as the formal account of how one system recognizes another as a "thing," a persistent entity—that is, the FEP is about what we mean by "thingness" (Fields et al. 2022). At the same time, the FEP shows that to be a "thing," a system must act; it must exhibit a basic form of agency—that is, it must manage information flow in a particular way (i.e., maximizing the evidence for its generative model). Thus, for Markov blanketed systems, being a "thing," acting, and exchanging information in a specific manner become essentially one and the same (Hohwy 2016, 281).

Furthermore, this particular form of agency should be viewed as existing along a spectrum: the complexity and temporal depth of self-evidencing depend on the nature of the Markov blanket and the generative models it involves and supports—Fields et al. (2022) and Friston (2019) distinguish between trivial and nontrivial agents. This gives us a highly flexible and physics-based notion of agency that extends to artifacts, provided they possess a genuine Markov blanket. Consequently, from the FEP perspective, design can be seen as the creation, assembly, and development of Markov blankets. We could also say that the Markov blanket embodies an artifact's "nature," while its generative model embodies its "function," and they each evolve on vastly different timescales. More precisely, the creation of an artifact is a cooperative act: the artifact is shaped not only by the designer (or group of designers) but also by the user (or group of users).

This chapter will focus on the first part of the argument. The next chapter will focus on the second.

3.1 NICHE CONSTRUCTION: ACTIVE INFERENCE APPLIED TO CULTURAL AND SOCIAL PROCESSES

Why did humans, unlike other mammals and primates, develop highly specific cognitive abilities that led to the emergence of complex symbolic communication, vast knowledge, and advanced technologies? Evolution and genetics alone cannot fully answer this question. One issue is time: from an evolutionary perspective, the 6 to 8 million years that separate early genus *Homo* from other primates may seem short for the emergence of complex cognitive and behavioral traits. Evolutionary and genetic processes operate gradually, and such a timeframe might be insufficient for these capabilities to develop solely through genetic selection. A plausible hypothesis is that a different factor made the real difference: the cumulative nature of cultural and technological transmission, also known as the ratchet effect (Tomasello 1993, 1999). This refers to the human ability to transmit knowledge and creative innovations within a group while modifying and improving them over time—essentially, allowing knowledge to evolve. According to Tomasello (1999), the ratchet effect was made possible by a uniquely human cognitive ability that researchers have not observed in other primates: the capacity to recognize and understand the behavior of conspecifics in terms of intentional states. In other words, humans can adopt other humans' perspective, a cognitive skill often referred to as mentalizing or mentalization (see chapter 6). This distinct form of social imagination laid the foundation for human sociogenesis, giving rise to language, rituals, artifacts, and institutions (Tomasello 2021).

From the perspective of active inference, the ability of agents to synchronize their respective generative models can be seen as the condition of possibility of mentalizing. A key reference is the model by Friston and Frith (2015b), which illustrates this process using the example of birdsong. Their work demonstrates how two active inference agents can establish communication, conceptualized here as *generalized synchrony* between generative models. According to Friston and Frith (2015b), when two active inference agents have sufficiently similar generative models, their mutual interaction enables them to predict their own sensory states in relation to the behavior of the other agent. As a result, their dynamics become synchronized, meaning that knowing the state of one enables the prediction of the other's states. In other terms, (1) agents start with similar generative models; (2) they have a shared structure but may have different initial states or parameter values; (3) through interaction, their predictions influence each other; (4) they adjust their internal states based on reciprocal feedback; and (5) generalized synchrony emerges, aligning their internal dynamics. This does not mean that the two agents share a generative model. Rather, they coordinate their respective generative models, which in turn enables the synchronization of their behaviors. (While the simulations and formal models in Friston & Frith 2015b deal with continuous variables, the underlying principle of generalized synchrony—mutual predictability through shared generative models—does not require continuity per se.)

Unlike Kaufmann et al. (2021), I argue that Friston and Frith's (2015b) generalized synchronization model provides the essential starting point for defining collective intelligence within the framework of active inference. This model is embodied, as synchronization arises from bodily interactions; endogenous, since it emerges from the intrinsic relationships between agents rather than being imposed by external rules or parameters; and scalable, allowing it to extend across different levels of organization. From this model, a general principle can be drawn: *when two or more active inference agents interact, their generative models tend to converge*—though the degree of convergence varies—*without explicit behavioral rules*. This follows directly from the FEP. However, as said, synchronization is merely a condition of possibility, necessary but not sufficient for understanding the so-called second-person perspective and highly complex social phenomena (more on this in chapter 6). (It is also very important to highlight the connection between synchrony and self-organization, as well as the phenomenon of interpersonal synchrony—both verbal and nonverbal. This is a central aspect of psychotherapy; see Koole & Tschacher 2016.)

The relevance of Friston and Frith's (2015b) model is also confirmed by Heins et al. (2024), who use active inference to explain the collective motion of a group, such as a flock of birds. In the simulations, the agents share a structurally similar generative model, specifically designed to capture collective movement through active inference. Each agent encodes the local distance between itself and its neighbors across multiple sensory sectors. However, while the overall structure of their generative models remains consistent, variations arise due to differences in sensory precision, responses to perturbations, and parameter tuning. The generative models in these simulations are based on a stochastic differential equation framework, allowing each agent to maintain a probabilistic representation of its distance to surrounding agents. To investigate behavioral adaptation, the study systematically varied key parameters such as sensory smoothness and velocity fluctuations. From this, a form of generalized synchrony gradually emerges—agents coordinate and move together harmoniously. Although agents do not share identical internal states, their predictive coordination enables them to align their behaviors based on mutual inference. This synchronization occurs without explicit communication, relying instead on shared inference mechanisms. It is not absolute—differences in sensory precision and adaptive learning introduce variability, making the coordination dynamic and context-dependent (see also Vasil et al. 2020).

It is important to note the role of the informed individuals in Heins et al. (2024). These individuals are agents that have an augmented generative model, allowing them to represent an additional latent variable: the distance to a fixed external target. They represent the leaders of the group. Unlike uninformed agents, who only track distances (i.e., distances to neighboring agents), informed individuals also infer and act on their relative position to a predefined goal. The results of the simulations (Heins et al. 2024, 8–9) show that increasing the proportion of informed agents in the group initially enhances the group's ability to reach the target, but only up to a certain point. Beyond a certain threshold, further increases do not necessarily improve or may even reduce collective accuracy, that is, the probability of reaching the target.

Let us delve into this more. Heins et al.'s (2024) analysis provides valuable insights into the role of leadership in groups from an active inference perspective. Their claim is that, as the proportion of informed agents increases, the group becomes better at reaching a common target because informed agents guide the movement through social interactions and information propagation. However, as said, this improvement holds only up to a certain threshold. If too few agents are informed, their influence is weak, and the group struggles to coordinate. If too many are informed, the group becomes overly sensitive to target noise, reducing efficiency and potentially causing fragmentation; that is, each informed agent independently navigates to the target. The optimal balance is when there are enough informed agents to guide the group while still maintaining social cohesion, ensuring the best accuracy in reaching the target (i.e., controlling information propagation). The threshold at which increasing the proportion of informed agents stops improving accuracy depends on three factors: (1) social precision, that is, how much uninformed agents rely on social cues given by informed agents; (2) target precision, that is, how accurately informed agents perceive or model the target (if target precision is too low, informed agents struggle to move correctly toward the target); and (3) group size, that is, larger groups require more informed agents to effectively guide movement, but if too many agents are informed, social coordination weakens (as said, in smaller groups, a lower proportion of informed agents may be sufficient, but too many informed agents can still lead to oversensitivity and fragmentation). Therefore, the balance is essential. I am convinced that this is also a very important line of research for understanding the dynamics of technological development (more on this later).

Returning to the previous point, synchrony is crucial for the formation of a social group, but it alone is not enough. According to Veissière et al. (2020; see also Veissière 2016), cultural affordances also play a crucial role. In very basic terms, an affordance is an opportunity for action provided by the environment—"a relationship between the properties of an object and the capabilities of the agent" (Norman 2013, 11). Affordances are not inherent in the environment itself but depend on the predictions and knowledge of the individual engaging with it. Therefore, affordances "are not simply static features of the environment, independent of the presence and engagement of an agent, nor are they states of the cognitive agent alone" (Veissière et al. 2020, 7). In fact, affordances emerge from the interaction between these two elements. From the perspective of active inference, affordance can be understood as a stable set of highly precise predictions about the environment and the possible actions it affords (Badcock et al. 2022; Bruineberg et al. 2018; Constant et al. 2018; Ramstead et al. 2016; Ramstead et al. 2018; Ramstead et al. 2019a). For instance, a chair affords sitting, but this affordance is only accessible if an individual knows how chairs are used in their cultural context. Similarly, more abstract cultural affordances, such as symbolic gestures or social roles, are learned through participation in social life (e.g., rituals)—affordances are always linked to signifiers. Affordances define what actions are possible, while signifiers indicate where and how those actions should happen (Norman 2013, 14). Hence, "the relational space of possibilities between agents and their environments constitutes an ecological niche" (Veissière et al. 2020, 7). Agents and their econiche "become attuned to each other, as the result of their history of co-adaptive interactions" (Veissière et al. 2020, 7) (figure 3.1). (For the role of affordances in human cognition evolution, see also Pezzulo & Cisek 2016.)

According to Veissière et al. (2020), active inference agents sharing the same econiche gradually develop common regimes of expectations—structured sets of predefined predictions that shape behaviors, social practices, and norms. Regimes of expectations are also connected to regimes of attention, that is, collective precision optimization practices. These highlight specific context-appropriate actions or perceptions, making them particularly salient, and guide individuals in learning how to navigate the social norms and resources of their local cultural environment. Among these resources, the most important are the epistemic tools that offer information deemed relevant and trustworthy (Constant et al. 2019b; Ramstead et al. 2016; Rendell et al. 2011; Veissière et al. 2020). In other words, regimes of attention refer to the structured ways in which individuals focus on specific elements of their environment, guided by cultural affordances and regimes of expectations. These attentional regimes determine what is seen as salient or important in a given context. Salience, in more technical terms, refers to the expected information gain or the degree of uncertainty resolved in that context (Friston 2010a, 2013; Parr et al. 2022).

It is important to stress that cultural affordances and regimes of attention cannot merely reproduce imitation or convergence around pre-constructed

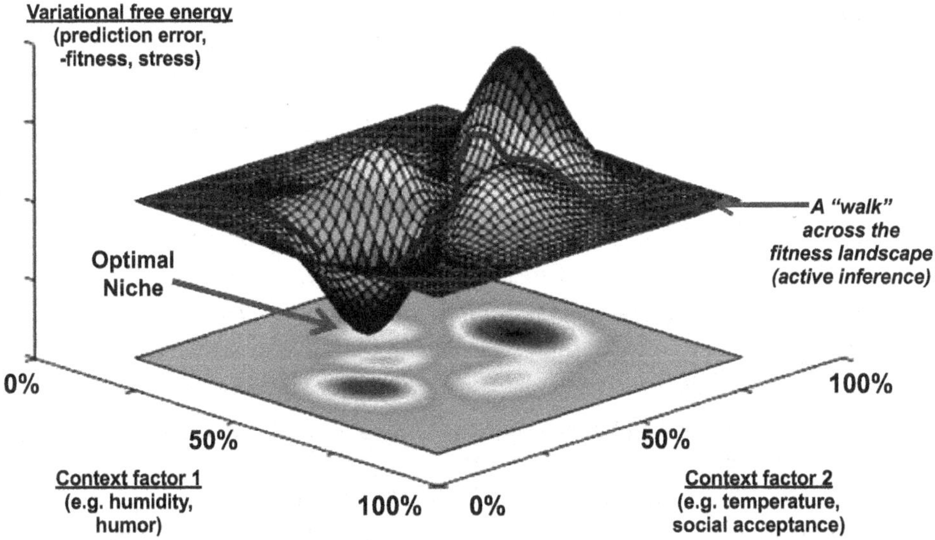

Figure 3.1
Active inference as gradient descent in a free-energy landscape. This figure depicts organisms as agents minimizing variational free energy by navigating a landscape of environmental niches. Action (niche exploration and construction) modifies sensory input, reshaping the fit between external states and internal generative models. Low points in the landscape correspond to econiches where prediction error is minimized and homeostatic equilibrium is achieved. The vertical axis represents free energy (local prediction error), while the horizontal plane spans possible environmental conditions. Through cycles of perception and action, organisms perform approximate Bayesian inference, dynamically adjusting their internal models to inhabit niches of maximal model fitness and thermodynamic stability. Source: P. Badcock, M. Ramstead, Z. Sheikhbahaee, and A. Constant, eds., 2022, *Applying the Free-Energy Principle to Complex Adaptive Systems* (Basel: MDPI).

systems of expectations (this would trap agents in the dark room problem mentioned in chapter 1). They must also create areas within the environment where uncertainty increases under control (and this also allows the ratchet effect). This is consistent with the nature of an active inference agent, which prioritizes actions that maximize epistemic value—that is, actions that are most salient or resolve the greatest uncertainty. What I mean is that *for active inference agents, minimizing uncertainty is a value in itself*—the active inference agent can tolerate bubbles of controlled uncertainty only if they serve to reduce greater uncertainty in the long run. This also means the active inference agent seeks out areas of the environment with the highest uncertainty—it is *inherently* driven by the search for uncertainty in order to minimize it. In other words, optimizing the pragmatic value of an action (its expected utility) is conditional on optimizing the epistemic value of that action. When I enter a dimly lit kitchen to grab a midnight snack, the first thing I do is not to go straight to the pantry and grab the snack (pragmatic value), but to turn on the light (epistemic value) (Veissière et al. 2020, 13). The search for salience precedes and conditions the practical reward.

Now, the central thesis of this survey is that these efforts to extend the FEP and active inference into the social sciences have overlooked technology's active role and significance. There is a lack of a more systematic exploration of the relationship between social affordances and technology. Yet the relevance of this aspect is clear: "Epistemic affordances are encoded by—or installed in—the environment, as repeated physical actions leave traces that change the structure of the developmental niche in ways that influence agents' expectations" (Veissière et al. 2020, 14). Cultural affordances arise from the transformation of the physical environment and the expectations of the agents living within it. The importance of technology is also emphasized by Clark (2016, 276): what truly characterizes human beings, he writes, is "our abilities of temporally co-coordinated social interaction and our ability to construct artifacts, and design environments." To clarify this further, Clark refers to language, considering it the fundamental model for any artifact.

From a classic active inference perspective, an artifact is an environmental arrangement that reshapes EFE for a given class of agents and tasks. It offers epistemic affordances—opportunities to reduce uncertainty via more informative observations—and pragmatic affordances—opportunities to reduce risk or ambiguity, or to expand the policy space so more effective (or entirely new) actions become possible (Constant et al. 2019b). This means that artifacts do not possess agency of their own, nor do they have autonomy. They depend on an agent's generative model and preferences and remain merely potential until the human agent actually realizes them. Defining an artifact only as a bundle of epistemic and pragmatic affordances, then, risks sidelining artifact's agency.

From this perspective, I believe that deeper engagement with the technology studies literature could better integrate the FEP into the social sciences. I argue that without recognizing how artifacts not only participate in but also actively shape inferential processes, any application of the FEP to social phenomena will remain fundamentally incomplete. Too often, studies transplanting the FEP into cultural and social contexts ignore the fact that technologies act to construct the very networks through which agency is distributed. For this reason, ANT's challenge to classical sociology can offer a powerful corrective to these studies. By showing how social order emerges from heterogeneous networks of human and nonhuman actors, ANT compels us to see artifacts as central agents in the delegation and execution of inferential capacities.

Consider autonomous vehicles: their navigation relies on GPS, infrastructure, regulations, software, passengers, weather, and insurance. No single element drives them; functionality emerges from their interaction. Likewise, scientific discoveries depend on equipment, funding, methodology, publications, and materials. According to ANT, agency resides not in individuals but across the entire research network (Akrich & Latour 1992; Callon 1989; Latour 1988, 1999, 2007; Law 1987, 2002, 2008). The social disappears as a "thing" and becomes a process, an association, or—in Latour's terms—a collective composed of processes of continuous hybridization of humans and nonhumans. Indeed, "scallops make the fisherman do things just as nets placed in the ocean lure the scallops into attaching themselves to the nets and just as data collectors bring together fishermen and scallops in oceanography" (Latour 2007, 107). The point is that "no one has ever seen a technique, and no one has ever seen a human. *We only see assemblies, crises, disputes, inventions, compromises, substitutions, translations, and orderings that get more and more complicated and engage more and more elements*" (my emphasis).

In the following sections, I will show how the FEP and active inference furnish us with exceptionally refined and precise physics-based conceptual tools to justify the attribution of agency to artifacts. This, in my view, constitutes the FEP's principal contribution to technology studies. And it also enables us to reinterpret and strengthen some of the key concepts in this field.

3.2 MACHINE AGENCY AND THE FEP

Let us return to our initial question: Why do we turn to active inference and the FEP to understand technology? Is it just a matter of repackaging familiar insights in new jargon? As the name suggests, active inference is first and foremost about agency. Now, the question of machine agency is one of the central issues in the philosophy of technology and design, as well as in the sociology and psychology of technology (Mattingly & Cibralic 2025). My claim here is that the FEP and active inference offer an elegant theoretical framework that not only justifies attributing agency to artifacts but also helps us rethink key concepts in technology studies, such as technical mediation and symmetrical ontology, in a more rigorous and meaningful way. In other words, the FEP and active inference provide us with a single, straightforward, physics-based conceptual framework—no additional principles required—for explaining not only agency itself, but also the continuity and distinctions among different forms of agency. Considering the agency of artifacts and the continuity among different forms of agency—across various types of artifacts and between artifacts and humans or animals—can significantly improve design decisions.

An artifact can be defined as any material product of human action, from straightforward items like hammers, eyeglasses, and notebooks to intricate, composite systems such as brain–computer interfaces, skyscrapers, and satellite networks. This idea echoes Aristotle's distinction between things brought into being by human agency and those arising through natural forces (Aristotle 1984). Artifacts are created intentionally by people (designers, inventors), whereas other objects and structures form independently in nature. Under this definition, we also count less conventional artifacts—perfumes, foods, songs, films, algorithms, and computer games—as human-made products. Heersmink (2021) presents a detailed taxonomy of artifacts, distinguishing four types: (1) embodied artifacts, (2) perceptual artifacts, (3) cognitive artifacts, and (4) affective artifacts.

Artifacts are inherently "for something," meaning they serve specific "functions" (Houkes & Vermaas 2010). Typically, an artifact's function is defined by the effect it produces in the world for its user: a hammer drives nails into surfaces, eyeglasses sharpen our vision, maps help us navigate, and guitars enable us to make music and express emotion. In most cases, these tools are deliberately designed and manufactured with a particular purpose in mind—what we might call their "proper functions." Yet we also often repurpose artifacts in ways their designers never intended. A hammer can become a weapon, eyeglasses a diadem, a map a placemat, or a guitar a decorative piece. In short, artifacts are functional entities—whether intentionally crafted or creatively repurposed—that produce effects in the world for their users.

Questions about whether artifacts deserve recognition as autonomous agents in social change—or whether agency belongs exclusively to humans—sit at the heart of technology studies yet remain unresolved. Many scholars argue that what looks like an artifact's "property" or power really boils down to human choices and interpretations, leaving no need to ascribe genuine agency to nonhuman things (Brey 2005, 2006; Heersmink 2017, 2021; Hornborg 2017; Kallinikos 2002; Mitcham 2013). In contrast, a large body of literature clearly demonstrates how a technology's design can significantly influence behaviors and drive social change. Latour (1992), for instance, shows how everyday objects like seat belts and hotel keys can nudge users toward specific behaviors. Take the hotel key with its large, heavy fob—it is designed so guests are more likely to return it to the front desk when they leave their room. Sclove (1995) observes that modern sofas with two or three separate cushions carve out individual seating areas, reflecting and reinforcing Western values of privacy and personal space—unlike, say, Japanese futon sofa beds. Winner (1980) argues that nuclear power plants inherently demand centralized, hierarchical management to operate safely, in contrast to technologies like solar power, which can function in a more egalitarian framework. Thus, the design of these technologies effectively dictates the social structures required for their use.

From a philosophical standpoint, there are essentially two positions (with varying degrees of nuance): (1) naïve realism, which assigns autonomy and agency to artifacts and, consequently, links those artifacts to certain social changes, and (2) social constructivism, which holds that it is the context—society itself—that ascribes those qualities and, therefore, that agency to the artifacts. More precisely, we can identify two extreme positions: (1) the strong view, exemplified by Latour and ANT, which develops a symmetrical or "flat" ontology in which agency is distributed equally among humans and nonhumans, with no ontological or epistemological distinction between different types of agency, and (2) the weak view, represented by postphenomenology (Ihde 1990; Verbeek 2005, 2008), the work of ecological psychologists (e.g., Gibson 2014), classical phenomenology (Heidegger 1927), and anthropology (Gell 1998; Ingold 2013), which maintains an ontological and epistemological distinction between human agency and nonhuman agency, so that nonhuman agency depends on human agency—without dissolving technological agency entirely into its social context.

Now, my thesis is that the FEP offers a simple theoretical model for addressing the question of artifact agency—striking a balance between the two extreme positions just mentioned. By grounding agency in principles from physics and information theory, the FEP gives us good reasons to ascribe agency to artifacts without sharply distinguishing it from the agency of humans, animals, or other living beings. If we accept that agency is essentially self-evidencing and emerges simply from the presence of a Markov blanket, then there is no reason we cannot ascribe agency to machines as well, with varying degrees of autonomy and complexity. A simple route to arguing that artifacts are agents from the FEP is that there is a Markov blanket coupling an agent and an artifact, and this means that we can treat both as agents by simply inverting the labels "active" and "sensory."

3.2.1 A MODEL OF MACHINE AGENCY

There are numerous theories of agency, most of which use human agency as their model and therefore often rely on mentalistic concepts rooted in folk psychology (Anscombe 1957; Davidson 1963; Dennett 1987, 2003; Dretske 1988; Frankfurt 1978). This perspective, as Mattingly and Cibralic (2025) point out, implicitly downgrades machine agency—treating it as a less rich, autonomous, and complex subcategory of human agency. If a theory defines agency in terms of mental states (e.g., intention, deliberation, reasons, self-reflection) and treats those states as paradigmatically human or as requiring human-like consciousness, then non-human systems can only count as agents by analogy or as mere approximations. As a result, machine agency is structurally classified as derivative, thinner, and less autonomous, because it is evaluated against criteria it is assumed not to fully meet.

In my view, the FEP allows us to move beyond this impasse by offering a nonanthropocentric conception of agency based on "a particular physics" (Friston 2019), namely, "Bayesian mechanics" (Ramstead et al. 2023).

Alongside other cornerstones of mathematical physics—for instance, variational principles such as the principle of stationary action or the principle of maximum entropy—the FEP underpins a new class of mechanics or mechanical theories (much as the principle of stationary action gives rise to classical mechanics, or the principle of maximum entropy yields statistical mechanics). This new mechanics has been called Bayesian mechanics.

It provides a formalism in which Markov-blanketed dynamical systems can be described *as if* their internal-state dynamics implement variational (approximate) Bayesian inference. In some worked examples, Sakthivadivel (2022c) shows that classical mechanics arises as a special case of Bayesian mechanics, where physical systems minimize surprisal in their trajectories. Classical trajectories are treated as maximum-probability (least-surprisal) paths under a maximum-caliber/path-entropy formulation; in the limit of vanishing uncertainty, probability concentrates on the classical path. In other words, classical paths are the most probable paths of a distribution over all possible paths. With "infinite certainty" (no fluctuations), the result is deterministic trajectories. The mode-matching (stationary states) / mode-tracking (dynamic states) / path-tracking (inference over complete trajectories) taxonomy characterizes increasingly rich approximations to this inferential description (Sakthivadivel 2022c). Interpreting forces as blanket-mediated couplings (e.g., via force measurements as sensory states) is a modelling move that illustrates how classical dynamics can be embedded in the Bayesian-mechanical framework. This approach unifies classical physics, Bayesian inference, and stochastic thermodynamics through a concrete example of self-organizing systems.

This physics-based approach lets us reconsider classical philosophical concepts like belief (i.e., the agent's encoded probability distributions over external states), desires and goals (i.e., preferences or preferred states), and intentionality (i.e., an objective function, VFE and EFE, that the system implicitly minimizes) outside a purely psychological framework, grounding them instead in statistical physics and information theory (i.e., in the equivalence of their equations). Furthermore, the FEP and active inference rightly highlight an often-overlooked aspect of agency theories: the critical role of attention.

From the FEP point of view, agency is essentially self-evidencing and depends directly on the existence of a *physically meaningful* Markov blanket (Kirchhoff et al. 2018, 2). It is important to stress that a Markov blanket is not merely a conceptual sticker you can slap on any arbitrary partition. As said, Sakthivadivel (2022a, 2022b, 2022c) demonstrates that—even in the "weak" sense—blankets only arise when internal interactions vastly outweigh external ones, typically in large systems operating on multiple, clearly separated timescales. Without genuine sparsity and distinct timescale gaps, the blanket simply disappears. Therefore, Markov blankets are "rare in simple systems *but ubiquitous in complex systems*" (Sakthivadivel 2022a, 16; my emphasis). Beck and Ramstead (2025) emphasize that identifying a blanket is neither trivial nor arbitrary. In practice, you must perform systematic, often computationally intensive checks to ensure that conditioning on the blanket truly insulates internal states from external influences.

A critic might object that the mere existence of a Markov blanket does not, by itself, amount to agency—you still need a generative model. On this view, a Markov blanket is "just" a statistical partition. I agree that a blanket, taken purely in the abstract, is not sufficient to define agency. But two points matter here.

First, in the active inference tradition, the Markov blanket is not an optional add-on to a generative model; it is the condition of possibility for specifying one in a physically meaningful way. A generative model, in this setting, is not merely a formal object we write down—it is a model that is realized by a system whose internal states are informationally insulated from external states except through a boundary (the blanket). Without a stable pattern of conditional independencies that separates internal from external states, there is no well-posed sense in which internal states could model, track, or be about external causes at all—so a generative model would be meaningless.

Second, the FEP is best understood as *physics with boundaries*, and both terms are essential. What does the "boundary" amount to? It amounts to a concrete organization of information flow: sensory and active states mediate the only systematic coupling between internal and external states. When such a bounded system exhibits stable nonequilibrium dynamics that preserve its organization over time, the blanket is not merely a statistical convenience—it is a physically maintained interface (this does not mean that the boundary must be material or concrete). Under these conditions, talk of a "generative model" becomes appropriate because the system's internal dynamics can be treated as encoding and updating probability distributions over external states relative to that boundary. In other words, once you have a physically meaningful Markov blanket plus sustained self-maintenance under uncertainty, you have the minimal substrate that makes an inferential description *intelligible*.

So, a Markov blanket understood purely as a static statistical partition is not sufficient for agency, and it does not automatically entail a rich, explicit generative model in the way cognitive scientists often mean. My claim is instead this: in active inference, a physically instantiated Markov blanket—together with the system's continued maintenance of that boundary in a changing environment—provides the necessary basis on which an implicit generative model can be ascribed, and on which the system can be described as behaving *as if* it were performing Bayesian inference. A physically meaningful Markov blanket is not necessarily material; it is a physically realized, dynamically maintained boundary between internal and external states.

This point becomes even clearer in the path-based formulation of the FEP. In that setting, the primary objects of analysis are not explicit generative models defined over discrete variables, but continuous trajectories of a system evolving in state space under stochastic dynamics (Friston et al. 2023a). What matters fundamentally are the flows, constraints, and attractors that shape these trajectories over time. In order for such a description to make sense as a model of inference or self-organization, there must already be a principled way of distinguishing internal from external degrees of freedom and of specifying how they are coupled along entire paths. The Markov blanket provides precisely this structural condition: it defines which components of the system participate in internal dynamics, which belong to the environment, and which mediate their interaction across time. Without a physically instantiated blanket that stabilizes this partition along trajectories, there is no coherent basis for interpreting the resulting path-wise dynamics as minimizing any free energy functional. In this sense, the generative model does not precede the blanket in the path-based formulation; rather, it emerges as an inferential interpretation of bounded dynamical flows, making the Markov blanket not merely compatible with, but foundational to, the path-based approach.

Therefore, a *physically meaningful* Markov blanket is essential for agency. Much of the FEP literature shows that self-evidencing is an emergent property of any random dynamical system endowed with a *physically meaningful* Markov blanket. However, this does not mean there is only one kind of agency; agency is a spectrum that ranges from a very basic and reactive form of agency to full agency including complex activities such as planning and counterfactual reasoning. The depth of an agent's agency depends on the architecture of its Markov blanket and the complexity of the associated generative models and prior beliefs. Full agency is not the mere execution of physical movements, but the existence of probabilistic beliefs about one's own actions—beliefs that are formally embedded in the agent's generative model and optimized via free-energy minimization (Friston et al. 2013).

If we adopt Friston et al.'s (2023a) particle taxonomy, then even a rock has a minimal level of agency, albeit only internal. A rock does not intermingle with or dissipate into its surroundings—at least not on a timescale over which we can still reliably reidentify it as the same rock. The rock exists as a rock (that is, as the particular thing it is) because it is separated from the rest of the system in a specific way. A rock is an inert particle that lacks active states. It still has a Markov blanket and a generative model composed by a dynamic prior over external causes (the physical laws governing what lies outside and at the boundary of the rock—continuum mechanics, gravity, heat diffusion, and other characteristics) and a sensory likelihood that links those causes to signals at the blanket (what happens at the surface) and, indirectly, to internal states. The rock's internal states engage in self-evidencing: they update by descending the gradient of VFE given what impinges on the blanket. However, as an inert particle, the rock has no active states; it cannot use actions to make its predictions come true in the world, so inference remains internal; we can't see their behavior. In short: updates (priors/beliefs) occur only internally; the world and sensors are not manipulated. For this reason we can only hypothesize the behavior of the internal states. Here, the Markov blanket establishes only a minimal and reactive self-evidencing.

A black hole can be modeled as a particle in the formal sense of Friston et al. (2023a), that is, as a system individuated by a partition into internal states, external states, and boundary states. However, the blanket does not exhibit a sensory–active factorization in the agentive sense, but consists of a restricted set of macroscopic boundary variables—specifically mass, charge, angular momentum, and associated energy and entropy fluxes—which, at an appropriate coarse-graining, largely constrain what is observationally accessible about the internal degrees of freedom beyond the event horizon. These variables mediate the effective conditional dependencies between inaccessible internal states and the external environment, yielding a non-agentive or purely observational blanket. The black hole is a dissipative system, insofar as it participates in nonequilibrium exchanges with its surroundings through processes such as accretion and Hawking radiation. However, it remains inert in the sense of active inference: there is no separable channel by which internal states modulate boundary dynamics to reduce VFE and EFE. It is a coarse-graining choice: the rock preserves a dynamically effective interior, whereas under the no-hair description the black hole's interior degrees of freedom are effectively marginalized into boundary variables alone. Accordingly, it is not appropriate to model the black hole as updating a generative model, selecting policies, or maintaining a viability set. This case therefore illustrates that dissipation and inferential agency are orthogonal dimensions in the classification proposed by Friston et al. (2023a).

Conservative particles, by contrast, possess active states and are therefore capable of full self-evidencing. Strange particles exhibit even more complex self-evidencing; their main feature is that active states become hidden causes, and therefore internal states encode beliefs not only about external states but also about their own active states (and this implies a sense of selfhood, full agency, or autonomy) (Friston et al. 2023a, 51–53). In other terms, we model internal, external, and blanket states as probability densities over trajectories. We then show that—under the conservative regime—the autonomous states (internal and active) behave as if they obey a least-action principle: the Lagrangian of particular paths and the VFE share the same minima. Hence, minimizing the VFE yields least-action paths for internal states—and for active states in the conservative case (Friston et al. 2023a). Consequently, a system's degree of agency depends on the scale and structure of its Markov blankets, since that frame directly determines how much complexity (and therefore how much temporal depth of beliefs and actions) its generative model can sustain. With this in place, we can define different agent types in terms of the constraints that are enforced or encoded at the boundary.

From this perspective, following Beck and Ramstead (2025), the flame possesses a Markov blanket and self-evidences by continuously minimizing surprise relative to its environment: its surface acts as a statistical boundary that isolates the internal dynamics (exothermic reactions) from the external influences (cold-air flow, pressure variations) (Beck & Ramstead 2025, 14–15).

Now consider a hammer. What sets it apart from a rock? The hammer, like the rock, is inert and has no active states. What makes it different is its coupling with the user—a goal-driven agent with a particular Markov blanket and generative model, that is, a strange particle. The key point is that, together, the hammer and the user constitute a new Markov blanket with its own generative model: the human drives action through the hammer. The hammer must have specific physical properties that let it couple with the human user—properties set by the blueprint, that is, the designer's generative model. In tandem, *hammer plus user* constitute an agent endowed with self-evidencing capabilities and a complex generative model. This brings out an important point—philosophically as well—namely, the *generative capacity* of the Markov blanket: the fact that one or more Markov blankets (e.g., the human or a group of humans) can give rise to other Markov blankets (e.g., the human plus the hammer or a group of humans plus axes or nuclear weapons), that is, new systems endowed with a specific agency. This clearly shows the concept's flexibility and malleability. However, there are some conditions. A Markov blanket exists when a set of internal variables is separated from the external world by an interface of sensory and active states. Multiple local blankets can yield a higher-level Markov blanket only if the aggregate preserves this shielding: the union must present a single interface that fully mediates exchange with the rest of the world; internal couplings must dominate couplings to the outside; and the internal dynamics must be stable enough to support a (meta)stationary distribution. Under these conditions, the aggregate admits a coherent macro generative model, and it is meaningful to say a new Markov blanket has emerged.

The hammer case shows also that most technological artifacts cannot count as full agents. These artifacts clearly mediate human action and behavior, yet they do not maintain generative models that support counterfactual inference, nor can they regulate the precision of their own states in response to uncertainty. This is why the hammer needs the human being to become a properly acting artifact. Even more complex systems—such as automated doors—often fail in a subtler way: when they break down, they do not learn from the failure, nor do they revise expectations about their environment. On the present account, this is decisive. Full agency is not a matter of interaction or influence alone, but of participating in an ongoing inferential process that distinguishes expected from unexpected states and updates accordingly. By making these exclusions explicit, the FEP does not merely extend agency beyond the human; it restructures the ontology of artifacts, drawing principled boundaries that earlier STS approaches left deliberately open. Thus, attributing agency to an artifact depends not only on the kind of Markov blanket and generative model it instantiates, but also on *how the artifact is embedded in and contributes to an encompassing inferential dynamic* composed by humans and nonhumans.

This also implies that some artifacts may achieve a high degree of agency as conservative particles. A useful illustration is a WhatsApp client process running on a smartphone. To avoid confusion, the relevant unit of analysis must be specified explicitly: the system is the app process, while its environment includes the user, the operating system, the device hardware, and the networked infrastructure with which the process interacts.

On this description, sensory states should be understood not as "inputs" in general, but as state variables at the interface that mediate the influence of external states on the process. Concretely, these include (for example) the values of user interface event buffers (touch/keyboard events), network socket/buffer states (incoming packets as represented in local buffers), and relevant operative system-provided status variables (e.g., connectivity flags, permission states, and resource/constraint signals). Active states, likewise, are not just "outputs," but interface state variables through which the process can causally affect external states: for instance, the states of outbound network queues/sockets, pending notification requests as handed off to the operating system notification subsystem, audio/vibration requests as handed off to device actuators. In this sense, what matters for the Markov blanket is not the physical act of "rendering" or "sending," but the boundary variables whose evolution mediates the coupling between the process and its environment.

Internal states are then the process states that are not directly accessible to the environment except via those blanket variables: runtime latent states such as session keys and authentication state, message-queue and retry logic, compression/encryption states, learned or configured parameters that shape behavior, and the broader control state of the application. External states comprise the degrees of freedom that influence, and are influenced by, the process only through the sensory and active states: user intentions and actions (as realized in interaction patterns), operating system scheduling and memory management, hardware constraints, network conditions and endpoints (including server-side states), and other interacting processes. Outcomes such as a displayed notification or a successfully transmitted message are best treated as environmental consequences of the coupled dynamics, rather than as internal states of the app.

Given such a partition, one may model the client process as if it embodied a minimal generative model of its niche, in the following sense: the evolution of its internal and blanket states can be treated as approximately minimizing prediction-relevant or variational quantities under constraints imposed by its design (e.g., maintaining connectivity, delivering messages reliably under latency and packet loss, managing resource limits). This does not mean that the blanket, taken as a purely abstract statistical partition, is sufficient for agency; nor does it entail that the app literally "represents" the world in a psychologically loaded sense. The claim is rather that, once the blanket is physically realized by stable patterns of conditional dependence at the chosen scale, the app's behavior can be modeled—non-trivially and predictively—*as if* it were performing approximate Bayesian inference in order to maintain a coherent, persistent mode of operation within an uncertain environment.

However, we can also consider WhatsApp at another scale or level of description. In fact, WhatsApp is not just a single app on my smartphone. It is a planetary system running on millions of devices at the same time. From this perspective, we should view WhatsApp as a collection of groups of active inference agents of different kinds, for example, human users, devices, engineers, designers, software, online protocols, satellite networks, and other agents. We can identify different kinds of local synchronizations among active inference agents and, accordingly, different kinds of co-minimization processes of free energy. We can describe specific groups of active inference agents and their interactions (including sets of epistemic and practical affordances) and thus understand and model their behavior at the local level. In other words, the global ecosystem (WhatsApp) can be seen as a heterogeneous ensemble of agents (users, devices, servers, teams, policies) with local couplings and synchronizations. There are no single generative model nor unified preferences; users have their own—and often conflicting—goals. At that scale, there may be no single well-defined Markov blanket for WhatsApp as a whole. The app is thus not a single agent but a dissipative multi-agent structure with local free energy co-minimizations.

Interestingly, the idea of a Markov blanket and agency as self-evidencing aligns perfectly with the minimal conditions for attributing agency to machines established by Mattingly and Cibralic (2025). According to these authors, agency is not necessarily related to consciousness; agents only need "to gather information for the outside world, modify their internal states in response to that information, and then make changes in the world prompted by their new internal states" (Mattingly & Cibralic 2025, 59). Mattingly and Cibralic (2025) use the term "representation" to denote the predictions and objectives that the agent seeks to realize but which cannot be assimilated to the old psychological conception of mental states—because they are embodied and enacted. Moreover, in my view, the active inference perspective is also compatible with those accounts of machine agency that define agency in terms of goal-seeking behavior. For example, Carlsmith (2022) defines an agentic system as one that can use "a model of the world that represents causal relationships between actions/outputs/policies and outcomes" and that can "select actions/outputs/policies that lead to outcomes that rate highly according to various criteria (e.g., objectives)" (Carlsmith 2022, 9). Based on this definition, Carlsmith (2022) supports the attribution of agency to AI algorithms, but not to simpler artifacts like thermostats (whereas Mattingly & Cibralic 2025, by contrast, do attribute agency to thermostats). Other views of machine agency place greater emphasis on autonomy—specifically, the absence of human supervision—as seen in Shavit et al. (2023). The advantage that active inference offers is its ability to synthesize these perspectives in a non-anthropocentric way. Not only that—active inference also enables a significant conceptual leap. The key point is that, from an active inference perspective, agency is always co-agency, in the sense that a system never minimizes free energy in a vacuum, but always in relation to a context and to other systems that observe it and interact with it.

From this perspective, active inference and the FEP are closely linked to another relevant framework, TAME (Technological Approach to Mind Everywhere) (Levin 2022). One of TAME's key claims is that cognition does not operate on a binary or discrete logic, but on a gradual, continuous spectrum. Levin (2022) argues that this gradualist view is supported by evolutionary history, which shows a steady progression of cognitive complexity, and by bioengineering's ability to create chimeras across different "natural species," further blurring traditional boundaries. In other words, cognition is not limited to complex abilities like language or logical reasoning; it includes any information-processing activity that occurs between perceiving stimuli and acting on them—primarily the capacity to retain information from environmental inputs and, based on that, to form predictions about the environment itself. Moreover, cognition is not tied to any specific material or biological substrate. It is not confined to organisms with brains; it can occur in a variety of substrates, including single cells, plants, animal tissues, swarms, and artificial systems. Levin emphasizes the plasticity of cognition's material substrate—which is always in flux—as well as the plasticity of cognition itself.

This brings us to another important point: the conceptual distinction that the FEP draws among life, mind, and agency. Agency is essential for life and mind—yet the reverse does not hold. As Constant (2020; see the numerical example) shows, we can perfectly well imagine a system that minimizes free energy but nonetheless drifts inexorably toward death, or one that lacks a mind in the ordinary sense of the term. From this standpoint, the FEP does not endorse any form of panpsychism, since it remains agnostic about what is living versus nonliving and what is mental versus nonmental. Agency conceived as self-evidencing is not a sufficient condition for life or mind, even though life and mind do entail forms of self-evidencing. Minimizing free energy with incorrect priors or precisions can yield maladaptive behaviors that are out of sync with the environment and other agents. In other words, free energy minimization leads to life and mind only under the appropriate conditions.

Situating the agency debate in more concrete epistemological and ontological terms, I adopt an operational realism about Markov blankets: a blanket is real to the extent that, at a given spatial scale s and within a temporal window t, the variables designated as blanket states make the internal and external states approximately independent, and this separation remains robust under intervention and repeated measurement by an observer. This is neither a metaphysical essence nor mere bookkeeping; it is a claim about manipulation-resistant invariants that make agentive descriptions explanatorily and predictively fruitful. Because coarse-grainings are not unique, I treat blankets as equivalence classes MB (s, t) that share the same interventionist profile: several partitions may qualify, but not all do. This stance licenses talk of artifact agency without overclaiming a unique, scale-free boundary. The next chapter supplies the formal criteria and tests for Markov blanket detection, and then "inverts" them into design procedures, so that the ontological commitment made here is cashed out in operational practice.

Endorsing this view does not commit one to Markovian monism (Beni 2021). It is true: the map is not the territory. Still, in scientific practice all we have are models and data, and we explore model space to explain the data better and better. In this sense, the FEP works as a model-selection heuristic: it supplies an operational physico-formal constraint (free energy as an upper bound on negative log evidence, with Occam style complexity penalties) that helps discriminate among models. Yet that, by itself, does not license an ontological reading of our models: predictive success needs to be complemented by stronger criteria (cross-domain robustness, successful interventions, convergence across independent models). Any ontological commitment depends on one's account of reality/realism and requires further argument; that lies beyond the FEP. Conversely, saying that Markov blankets are merely arbitrary theoretical constructs is mistaken. They are not physical boundaries in the sense of material edges, but that does not make them mere abstractions. Markov blankets are genuine boundaries as statistical patterns: they have operational reality when they satisfy criteria of prediction, intervention, stability, and generalization. Recognizing them as such neither reifies them ontologically nor embraces naïve physicalism; it treats them as robust modeling constraints. We can also argue, at the same time, that the very plasticity of Markov blankets does not weaken but instead supports a non-essentialist and interactionist ontology (see below for the possible connections between active inference and object-oriented ontology).

Now, my argument is as follows. If artifacts exhibit more or less sophisticated agency depending on the architecture of their Markov blanket, then the design and the use of an artifact must be understood in two senses: (1) as engineering Markov blankets and associated generative models and (2) as a form of co-agency, in which the agency of the human and that of the artifact (or of multiple artifacts used to construct/use other artifacts) converge and synchronize to jointly minimize free energy. Before getting there, however, I will bolster the first part of my argument by demonstrating how the FEP—precisely because it justifies attributing agency to artifacts—helps us reframe and reinforce key concepts in technology studies.

3.2.2 UNDERSTANDING TECHNICAL MEDIATION: POSTPHENOMENOLOGY THROUGH THE LENS OF ACTIVE INFERENCE

Postphenomenology, a field rich in diverse interpretations and continuously expanding (Botin and Berling Hyams 2021; Liberati 2016; Prunotto 2022; Wellner 2020), defies a straightforward definition. In particular, it has taken shape as both a driving force behind and a result of the pragmatic and empirical turn in the philosophy of technology since the 1970s (Achterhuis 2001). Postphenomenology shares elements of classical phenomenology (particularly Husserl, Heidegger, and Merleau-Ponty) and pragmatism (particularly Dewey), made more applicable through science and technology studies (Bijker et al. 1993; Kroes & Meijers 2000; Rosenberg & Verbeek 2015). Its proponents give special attention to specific technologies, their design, and their empirical and practical impacts. They view technology not merely as a neutral, impartial instrument but as an integral mediator that affects human intentions and shapes the interaction between individuals and their environment. In the words of Ihde: technical mediation is "the phenomenological capacity [of an artifact] to bring into presence that which was previously undetected and even invisible, but precisely in this *difference* it also transforms the way in which the phenomenon may appear" (Ihde 1986, 49).

The postphenomenological notion of technical mediation attempts to overcome the dualism between subject and object, as well as subjectivism and objectivism. Verbeek reinterprets the Husserlian notion of intentionality as "directedness," which is no longer a relationship between humans and world, but a "fountain" from which, via technology, both subject and object emerge. As Verbeek explains: "Virtually all human perceptions and actions are mediated by technological devices, ranging from eyeglasses and television sets to cell phones and automobiles. These technological mediations do not so much take us to 'the things themselves' that classical phenomenology was longing for but rather help construct what is real to us" (Verbeek 2008, 14). Therefore, technology possesses autonomous agency, even though it ultimately derives from and depends on human intentionality.

Although postphenomenology has matured and influenced design, the idea of technological mediation is still too often treated as a metaphor or reduced to a superficial analysis (Possati & Aydin 2024). Moreover, postphenomenologists simply provide a description of the technological agency without delving into the reasons behind it.

Now, the FEP enables us to redescribe the concept of technical mediation without betraying—or even while reinforcing—the core insights of postphenomenology. To grasp this point, recall that the Markov blanket concept hinges on two interrelated aspects: (1) conditional independence and (2) mediated co-influence between internal and external states. Despite their formal independence, the two systems (the agent and the environment) ideally synchronize so that the agent becomes a statistical model of its environment, and, conversely, the environment becomes a statistical model of the agent (Kirchhoff et al. 2018, 4). This dynamic, active co-shaping of subjectivity and objectivity perfectly mirrors Verbeek's definition of technical mediation. The transformation of the environment (the birth of technology) necessarily entails a transformation of the very agent that reshaped it. The Markov blanket is an ongoing cycle in which influences flow bidirectionally—between internal and external states—mediated by the blanket. In short, "given enough time, agents will come to be the authors of the external states (i.e., environments) that reciprocate with predictable, uncertainty resolving sensory feedback of exactly the right sort to sustain cycles of self-evidencing" (Kirchhoff et al. 2018, 5). Technological mediation encompasses the entire cycle. Technology plays a dual role in shaping uncertainty. On the one hand, agents actively construct technologies that stabilize their environments, making sensory inputs more predictable and supporting ongoing cycles of self-evidencing. On the other hand, technological systems can also introduce new forms of uncertainty, reshaping the space of possible actions and interpretations. In this way, technology does not merely resolve uncertainty but reorganizes it, generating new adaptive demands and opportunities for inference and control. From this perspective, free energy minimization is always a hybrid process—humans and nonhumans jointly minimize it through shared agency and co-active inference.

One of postphenomenology's essential features is its taxonomy of technological mediations (Ihde 1990; Verbeek 2005), which includes the *embodiment* relationship—where technology integrates directly with the body and, together with it, shapes our engagement with the world (eyeglasses are a classic example); the *hermeneutic* relationship—where we must interpret technological outputs (as with MRI scans) to access knowledge about the world; the *alterity*, or otherness, relationship—where humans interact solely with technology while the world remains in the background (think human–robot interactions); the *background* relationship—where technology forms the ambient context in which we act without direct human–technology coupling, yet fully defines our relation to the world (e.g., air conditioners or windows); and the *cyborg* relationship—where the bond between human and technology becomes so deep it is symbiotic (such as brain implants). The effectiveness of the FEP is evident in that any form of technological mediation can be readily described in terms of Markov blankets and generative models—namely, as instances of co–free energy minimization in which each agent, by minimizing the free energy associated with its own internal states, also indirectly minimizes the free energy of the other agent, and vice versa. For example, the background relation can be modeled as a case of nested Markov blankets, where the two systems (the human and the artifact) remain autonomous even though they influence each other and refer to a collective of Markov blankets. The embodiment relation, by contrast, entails sharing a Markov blanket, but not necessarily a nested relationship.

The hermeneutic relation, in turn, involves a more complex dynamic in which the agents' generative models exhibit more temporal depth and hierarchical complexity. The FEP thus introduces both descriptive constraints and a unifying rationale for technical mediation. The rationale comes from the fact that the differentiation of artifact types and the connections among them all obey a single logic: the minimization of free energy.

3.2.3 AGENCY AND SYMMETRICAL ONTOLOGY: ACTOR NETWORK THEORY (ANT) AND ACTIVE INFERENCE

ANT adopts a macroscopic, sociological stance: it seeks to understand technology's constructive role within human groups—highlighting the hybrid nature of any groups made up of intertwined human and nonhuman actors, what Latour calls "collectives." Postphenomenology, by contrast, works at a microscopic, phenomenological level, directly examining the relationship between the human subject and technology. In ANT, the idea of artifact agency is connected to the concept of symmetrical ontology. Likewise, as we will see, active inference enables us to formalize and advance this concept in a meaningful way. ANT emerged from the need to renew social theory in light of developments in science and technology studies. It is not another iteration of poststructuralist deconstruction. On the contrary, its aim is to *reassemble the social*—to trace and make visible the processes through which human and nonhuman agents connect and collaborate in producing social reality. ANT is, above all, an empirical method. It does not begin with grand theoretical assumptions but instead follows the actual formation and reformation of techno-social groups (Cronon 1991; Latour 1993a, 1993b, 1993c, 1993d, 1994, 2007; McNeill 1976).

ANT's central methodological rule was formally introduced by Callon in 1986 and is known as the principle of generalized symmetry. The principle suggests that humans and nonhumans (such as technologies, objects, and natural elements) should be analyzed with equal importance from an ontological perspective. Instead of assuming that people are the only active or decisive agents, ANT argues that all components of the network—whether human (like individuals, groups) or nonhuman (like animals, plants, tools, buildings, or data)—play an active role in shaping interactions and outcomes. ANT is not the empty claim that objects do things instead of human actors: "It simply says that no science of the social can even begin if the question of who and what participates in the action is not first of all thoroughly explored, even though it might mean letting elements in which, for lack of a better term, we would call non-humans" (Latour 2007, 72). By granting both humans and nonhumans agency, ANT provides a more balanced view of how actions are co-produced within a network, showing that the behavior of a social system emerges from the interplay of all its parts. "We give ourselves a rule of the game not to change register when we pass from the technical aspects to the social aspects" (Callon 1986, 176). This principle is a methodological strategy "to remain agnostic concerning the question of what counts as an actor[1] in a given setting and course of events" (Blok et al. 2020, 8). To achieve true symmetry, concepts like nature and culture, or language and world, should no longer serve as explanatory principles; rather, they become the very issues requiring explanation. ANT's work is to show that behind these concepts there are actually assemblages, networks of interacting human and nonhuman agents.

There are, however, significant epistemological limitations to ANT. Latour's narrative elegance has undeniably helped make ANT both influential and engaging, but that very style can become an obstacle when you try to turn metaphor into method. ANT has played a crucial role in legitimizing the importance of objects and technologies in social dynamics. Yet on an epistemological level it remains open to criticism for its lack of conceptual clarity, excessive descriptivism, endless relativism (Collins & Yearley 1992), and absence of critical reflexivity (Whittle & Spicer 2008). Symmetric ontology risks becoming trivial; if everything is a network, then everything starts to look the same.

Active inference offers a way to justify symmetrical ontology—and, by extension, the ANT approach—overcoming these epistemological hurdles. By defining any system that sustains a Markov blanket as an agent, the FEP offers a seamless foundation for symmetrical ontology: humans, artifacts, and environments alike become instances of the same formal structure, distinguished only by the complexity of their Markov blankets and generative models. In this view, a person and a smartphone both "act" and "perceive" through their respective active and sensory states, and their interactions form a network of coupled Markov blankets and generative models. This turns the symmetrical ontology championed by ANT into a precise, computational framework: each agent is a node in a graph of inference, with edges representing dependency relations in the joint model. Consequently, agency is no longer a rhetorical attribution but a measurable property of any system with a Markov boundary, allowing us to simulate and analyze sociotechnical systems—the interactions between designers, devices, and users—on exactly the same terms.

This idea resonates deeply with the current active inference literature. For example, in Friston et al. (2024, 3), "a distributed network of intelligent systems, which interact frictionlessly in real time, and compose into emergent forms of intelligence at superordinate scales" is explicitly theorized. The nodes of such a distributed, interconnected ecosystem "may then be human users as well as human-designed artifacts that embody or implement forms of intelligence" (Friston et al. 2024, 3). Consequently, intelligence, or agency, should be conceived as a set of interconnected nodes that extends and propagates among agents that are both connected and nested within one another. In order to explore these "ecosystems of intelligence," we must "move beyond brains and embrace the active and nested characteristics of natural intelligence, as it occurs in living organisms and as it might be implemented in physical systems more generally" (Friston et al. 2024, 3). Moreover, Ororbia and Friston (2024), with their concept of "mortal computation," have shown that it is entirely plausible to envision an AI system based on adaptive active inference. What really matters is not the physical implementation (carbon or silicon), but the architecture of the Markov blanket and generative model.

An objection to our argument might be that it is ad hoc—that is, that we are inventing a new definition of agency as self-evidencing and coining neologisms specifically to justify attributing agency to machines (and this just because we want to be neo-materialists). However, that objection does not hold up. The FEP was not formulated to justify machine agency—it is, as noted, a purely mathematical framework. Moreover, the FEP gives us a clear definition of agency that is also sensitive to contextual differences (so agency is both a global characteristic of a certain type of dynamical system and something that manifests differently depending on context—and thus depending on the Markov blanket, the system's generative model, and the environment). Finally, the FEP aligns with our intuitive understanding of agency (i.e., intentional action toward an end). It is true that the FEP by itself does not include fundamental elements like subjectivity, historicity, or moral responsibility—but it does not preclude them either. Self-evidencing underpins every form of agency: a dynamical system with boundaries and an internal model of its environment minimizes the free energy of its internal states, which in turn drives it to act in the world—that is, to sample sensory data in accordance with that model. However, mere self-evidencing is not sufficient to develop the complex agency required for survival in highly intricate contexts.

From a strictly philosophical perspective, active inference and the FEP, in my view, could benefit greatly from a dialogue with object-oriented ontology (OOO)—a philosophical position that should not be confused with ANT or with new materialism, even though it is closely related to them (Harman 2016). In fact, the question "What is an object?" is crucial for active inference and the FEP; that is, it is their underlying philosophical issue (Friston 2019, 4–6).

For OOO, a good ontology should include every kind of object—from atoms to galaxies, from single-celled organisms to large multinationals, and even imaginary, nonexistent entities like Sherlock Holmes—while also providing a basic typology of objects and their qualities. Accordingly, OOO theorists adopt a very general definition of "object": an object is anything that cannot be reduced entirely to its component parts nor to the effects it has on other objects—"an object is more than its pieces and less than its effects" (Harman 2018, 53). This view follows not only from a commitment to a non-anthropocentric flat ontology, but also from a fourfold critique: of physicalism (all reality is physical or reducible to physics), smallism (all reality must be reducible to smaller and simpler components), anti-fictionalism (only actually existing things are real), and literalism (everything real must be translatable into clear, direct, literal propositions). New materialism (Barad 2007; Bennett 2010; DeLanda 2016) errs by reducing objects to their components, just as ANT errs by reducing them to their effects; in both cases, an object's existential autonomy and stability are erased. Following Husserl, Harman preserves the object's autonomy. First, he conceives the object as the simple correlate of a perceptual act. This correlate can be perceived in various ways or perspectives (sensual qualities), yet it remains the same—it remains stable (in active-inference terms: there is stability in the variables that make up its Markov blanket, which allows us to see it as more or less the same). Second, he conceives the object as something noncorrelated—that is, not relative to any perceptual act. This is, according to OOO, the real object and the real qualities of the object. This "inwardness" or "noumenal inwardness" (Harman 2018, 82) is not mysterious or unknowable; it is simply an ontological condition: the object's withdrawal from perceptual acts does not undermine its autonomy—it is its very condition. The sensual object and sensual qualities exist only as correlates of real objects and real qualities (whether human or nonhuman). The relations among these four terms (real objects, real qualities, sensual objects, sensual qualities) form the core of OOO.

A central aspect of OOO is the "mutual darkness" (Harman 2018, 12) of objects—the idea that objects remain "screened" from one another and never come into direct contact except through specific interfaces or mediators. Hence there is an internal depth to objects and an external interface—the Markov blanket "screens" the inside; the outside never fully accesses the system. The real object is an autonomous dynamical system that persists and typically has an effective boundary. The sensual object is the relational profile the object takes on in concrete interaction with another object; in active inference terms, it corresponds to the class of sensorimotor expectations and observational patterns that stabilize through the interface of their respective blankets. Sensual qualities are these expected/observed regularities and affordances; real qualities are the causal dispositions (the parameters/dynamics of the external generative process) that make them possible. The agent's posterior is the internal representation of that sensual profile from its side; it neither exhausts nor defines the profile on its own. By reinterpreting Kant's notion of the thing-in-itself through Heideggerian withdrawal, OOO strikes a perfect balance between immanence and transcendence. The theoretical advantage lies in moving beyond correlationism and anthropocentrism and opening up to a perspectival approach that avoids any reductionism. The FEP can translate these ideas into an operational formalism.

In any case, the crucial point is that the FEP allows us to move beyond a merely rhetorical and metaphorical formulation of machine agency. It provides us with a measurable and testable criterion for assessing the attribution of agency to machines. Agency, in the strong, nonmetaphorical sense implied by the FEP, is a rare and fragile dynamical regime, achieved only under demanding physical and informational conditions. While many systems—such as thermostats, mechanical regulators, or simple feedback controllers—can be described as having probabilistic boundaries that separate internal from external states, such systems do not thereby qualify as proper agents (they are at the lowest level of the self-evidencing scale, they are conservative particles). What really distinguishes agents is not only boundary maintenance, but also the ability to sustain temporally deep generative models, to modulate precision across multiple timescales, and to actively counter entropic drift through self-evidencing dynamics. Full agency, on this view, is neither ubiquitous nor guaranteed; it is costly, unstable, and often lost, as evidenced by cases such as biological organisms under anesthesia or severe brain injury, where physical boundaries persist but inferential agency collapses. This account sharply differentiates an inferential notion of agency from the liberal attributions common in ANT and postphenomenology, while remaining fully grounded in physics and information theory rather than anthropocentric notions of intention.

3.2.4 RETHINKING THE EXTENDED MIND WITH THE FREE ENERGY PRINCIPLE

This section is inherently connected to ongoing debates surrounding concepts such as distributed cognition (Hutchins 1995), the extended mind (Clark 2008, 2014; Clark & Chalmers 1998; Menary 2010), and Material Engagement Theory (MET) (Malafouris 2013). The purpose, however, is not to provide a detailed discussion of these positions. At the same time, the extended mind and MET are open to critique. For instance, proponents of the extended mind often rely on metaphors or focus too narrowly on specific cases (e.g., the famous example of Otto and Inga's notebook in Clark & Chalmers 1998), without identifying the underlying principles that govern the extension of the mind. *Why does the mind need to extend? What principles guide this extension? Do these principles vary by technology, society, or cultural tendencies, or are they universal?* Without clear answers to these questions, the concept of "extension" remains an evocative metaphor but little more than that. By contrast, MET, which shares key affinities with ANT, presents a more radical framework that helps to sidestep some of these limitations. According to MET, cognition does not occur solely within the isolated human brain but unfolds as a form of action in and on the world. Cognition is enacted—it emerges as a continuous process of interaction between organisms (human or animal), objects, artifacts, and environments. The mind, in this view, is not a thing but a process that exists in the intermediate space between brain, body, and environment (physical, social, and cultural). Nevertheless, MET also suffers from certain limitations. One major issue is its lack of rigor: it seems to suggest that cognition can extend to virtually anything. If, as MET posits, artifacts are not mere extensions but play an active role in cognition, how exactly does this influence occur? For example, in the case of pottery shaped by an artisan (as explored by Malafouris 2013), does the pottery itself "think"? Are we at risk of overvaluing materiality at the expense of mental processes? Are all material interactions equally significant for cognition?

Active inference and the FEP can provide both descriptive constraints and a unifying rationale to address these questions. To support this claim, I will draw on the argument developed in Clark (2022). For Clark, "the neuronal component of any putative extended organization is a biological brain whose core operating principle is to minimize long-term prediction error" (Clark 2022, 120). The problem is to understand the extension of the mind in terms of active inference. Central to the argument is the concept of the predictive body budget (Feldman Barrett 2017). This is a kind of "internal accounting that looks to our future bodily needs, addressing predictable shortfalls—such as inadequate blood sugar levels—well before they arise" (Clark 2022, 120). Just as a financial budget tracks future expenditure, a body budget "tracks and anticipates the use and replenishment of key resources for maintaining bodily life and functioning" (Clark 2022, 120). Building on this idea, Clark introduces the concept of the predictive brain budget, which is essentially its cognitivist counterpart—in building this second budget, our brains do not care whether key information is stored using biological memory or beyond, in notebooks, apps, and GPS systems; "what matters is just that the right information or operations become available as and when they are needed for the fluid control of behavior" (Clark 2022, 120). Epistemic and practical actions are deeply interconnected in keeping these two budgets in balance. From this, Clark concludes that "extended minds . . . thus emerge when ordinary predictive biological brains minimize expected future prediction error using the best means available" (Clark 2022, 123). He then adds: "knowledge-budgeting activities do not always imply the presence of extended cognitive machinery. But they do so whenever the sensorimotor loops involved in those activities have become sufficiently ingrained and well-practiced, and the relevant external resources are robustly available as and when such loops require them" (Clark 2022, 123). Therefore, Clark (2022) identifies two conditions for the extension of active inference: (1) a sensorimotor loop must repeat over time to become ingrained in memory and behavior, and (2) the necessary means (relevant external resources) must be available. However, in my opinion, this argument fails to explain the mechanism(s) through which extension occurs because it assumes what it should explain—namely, the construction of the artifact itself. But this is exactly the question I want to ask: *How does the artifact come into being? Why does the active inference agent need to extend itself through the artifact?*

My argument intends to develop Clark's thesis in order to better connect the extended mind approach and the FEP. I therefore propose two points:

- We can hypothesize a threshold beyond which constructing artifacts becomes necessary to further minimize EFE (i.e., to select policies that keep future sensory surprise bounded)—something that cannot be achieved (in terms of both computational complexity and bounded rationality) by using physiological means (current bodily and cognitive budgets). This context-dependent threshold emerges when resource constraints and precision limits make further improvements in inference without environmental scaffolding inefficient. At this stage, artifacts are created as external tools to support and integrate the agent's inferential processes.
- These artifacts are not arbitrarily constructed; rather, they are shaped by the designer's generative models—that is, by the individual or group who envisions and creates them. Their design reflects the designer's physical and cognitive constraints, as well as the predictions embedded in these models. As a result, a structural coupling naturally arises between the designer's generative models and the artifact's form and functions.

The threshold for artifact creation can be described as a policy choice between the following:

- *Policy 1:* Keep inferring with the existing model and body alone.
- *Policy 2:* Invest resources to create an artifact and then infer with the augmented system.

The agent then calculates the expected free energy (EFE) for each policy by balancing epistemic value (does the artifact reduce uncertainty, i.e., information gain?) and pragmatic cost (expected risk and resource expenditure relative to prior preferences and bodily/cognitive budgets). The threshold emerges when the epistemic benefit of creating and using the artifact outweighs its expected costs. The agent simply grants higher posterior probability (precision-weighted selection) to the policy that requires interaction with the artifact.

In more phenomenological terms, we can break this down as follows:

- *Phase 1: Break point:* The number of prediction errors is too high to manage, leading to an excessive need for the updates of the generative model. The agent experiences a breakdown in its ability to handle the environment as-is. The bodily and cognitive threshold has been crossed.
- *Phase 2: Formulating a generative model about the prediction errors (GM1):* To address this, the agent formulates a new generative model (GM1) to identify the causes of the prediction errors. Very simple example: the agent finds themselves in darkness and concludes that it is dark because the sun has set (absence of the sun = darkness). This new model provides a basic explanation of the observations and serves as the foundation for further action. Restructuring a generative model (or creating new models as an alternative explanation for new experiences, or for planning new actions) is sometimes known as structure learning (Smith et al. 2020b). Under radical constructivism, this is sometimes framed in terms of growing a mind (Tenenbaum et al. 2011).
- *Phase 3: Designing a solution (GM2):* Using GM1 as a starting point, the agent creates a second, more complex generative model (GM2). In the current example, the agent cannot simply change its expectations—it needs light to survive. The solution requires modifying the environment to create an "alternative sun," or a substitute source of light. The agent must imagine an *alternative* environment. As Manrique and Walker (2023) explain, innovative behavior relies on complex computational work: the creation of a controlled uncertainty space for the purpose of probing new hypotheses is only possible if the precision of prior beliefs is decreased. GM2 is built upon GM1 but introduces new priors and precisions, and concrete actionable steps. Therefore, GM2 incorporates policy evaluation in terms of EFE, enabling the agent to explore potential solutions and select the most effective one. After testing various policies, the agent identifies a successful strategy: lighting a fire. By performing a specific action (e.g., striking a flint), the agent can create fire and achieve the desired outcome: more light. This process can be called invention and, as we will see in the next chapter, involves creating a new effective boundary, a new coupled system that can be modeled with a Markov blanket.
- *Phase 4: Ensuring long-term stability (GM3):* While GM2 provides a solution, it is not sufficient. The agent now develops a third, even more

sophisticated generative model (GM3). This model focuses on controlling and maintaining the fire over the long term by identifying stable, repeatable policies (i.e., policies that can be communicated and transmitted to other members of the group). For example, the agent recognizes the need to stockpile wood to fuel the fire and to contain the flames to prevent accidental wildfires. GM3 is deeper and more temporally expansive than GM2, incorporating strategies for managing resources and ensuring safety.

From this angle, active inference aligns with a long philosophical tradition that treats the relationship between humans and technical objects as intrinsic. Artifacts are not external add-ons to human cognitive capacities beyond the limits of body; they're expressions of those capacities and dispositions in relation to the environment (Simondon 2005). For Sloterdijk (2011–2016), the human being is "spheric," meaning we build spheres—protective, technical, and symbolic spaces in which we grow, learn, and form identity while shielding ourselves from the uncertainty of the "outside." These spheres—from the womb to the family, school, the city, and today's vast digital habitats—serve as envelopes of meaning and protection from uncertainty. On this view, technologies are not merely extensions of mind; they are conditions of possibility for human development because they address the basic human need *to cope with uncertainty*.

3.3 CONCLUSION

This chapter has laid the groundwork for defining a design methodology based on the FEP. The central argument I have developed and defended is that the FEP provides a sound rationale for attributing—even minimally—a form of agency to artifacts and material culture and offers a sufficiently flexible foundational model for describing that agency, its limits, and its typology. I have further strengthened this thesis by showing how, through the FEP, we can reinterpret several key concepts in technology studies. This opens the door to a reformulation of the very concept of design in terms of creating Markov blankets and Markov blanket architectures.

4

DESIGN AS THE ENGINEERING OF MARKOV BLANKETS: A ROADMAP

This chapter develops the last part of the argument introduced at the beginning of chapter 3. The existence and type of Markov blanket are the key structural conditions under which a system admits an interpretation as a (minimal) self-evidencing agent—and this is also largely confirmed by the path-based formulation of the FEP (Friston et al. 2023a). It also follows that agency exists along a spectrum: both the complexity and the temporal depth of self-evidencing depend on the specific properties of the Markov blanket and the generative models it involves and supports. This yields a highly adaptable, physics-based concept of agency that can extend to artifacts, provided they possess a Markov blanket. This also means, from the FEP perspective, that design can be viewed as engineering Markov blankets. In this chapter, this notion of design is developed in two directions. First (in sections 4.1 and 4.2), the chapter seeks to establish the foundations of a design methodology based on Beck and Ramstead's (2025) algorithm to detect Markov blankets. Second (in section 4.3), it tackles the relationship between user and artifact in terms of synchronizing their generative models and co-minimizing free energy.

Let us go back to the question we asked at the beginning of this book: Why should a designer or engineer care about active inference and the FEP? What is the real advantage of having the formal model of an artifact as an active inference agent? Active inference and the FEP provide a normative design principle that unifies perception, decision-making, and control. In this framework, perceptual inference means reducing variational free energy (VFE) by updating the agent's beliefs; action selection and planning mean reducing expected free energy (EFE) over future horizons—favoring policies that make preferred outcomes likely while also reducing the kinds of uncertainty that matter for control. Therefore, when we formalize an artifact as an active inference agent—by specifying a generative model with likelihoods, dynamics, and preferences, and by stating the relevant design constraints—we obtain a Bayes-optimal (within-model) account of that artifact's behavior, given that generative model and the chosen variational approximation. In practice, the agent updates its beliefs via variational Bayes and selects policies by minimizing EFE, that is, maximizing expected model evidence (self-evidencing), according to an internal, checkable criterion grounded in the standard variational bound.

This formalization also builds in a particular kind of *sustainability*: preferences are expressed as distributions over observable outcomes, which implicitly define a set of viable states for the agent. By reducing EFE, the agent tends to avoid undesirable outcomes, limit observational ambiguity, and seek useful information when needed—thereby keeping itself within its viability set over time. In reinforcement learning, that sustainability-like behavior is not native to the basic objective. Reinforcement learning optimizes expected return; anything about staying within viable states must be added explicitly. In other terms, reinforcement learning can emulate sustainability if you engineer it in, but it is not guaranteed by default. Active inference, by contrast, makes viability-preserving, uncertainty-aware behavior a direct consequence of the core decision criterion.

Therefore, having an active inference formalization means you possess an explicit generative model of the artifact with preferences encoded as distributions over outcomes, a variational inference procedure to update beliefs, and a single decision rule: select policies that minimize EFE (the sum of risk, ambiguity, and epistemic value). This yields a clear normative criterion, traceability from design to behavior (every action justified by priors and likelihoods), built-in diagnostics for design and validation (EFE decompositions), compositionality via Markov blankets, and sustainability properties: the artifact tends to maintain itself within a viable set of states and seeks information when uncertainty limits control.

4.1 THE ACT OF DESIGN AS ENGINEERING STATISTICAL BOUNDARIES

I have shown that the FEP and active inference allow us to justify attributing agency to artifacts, giving us the conceptual means to strengthen and clarify some key concepts in technology studies. A direct consequence of these insights, in my view, is that design consists in engineering, stabilizing, and coupling Markov blankets. This is the starting point for developing a design methodology based on the FEP. Saying "design can be understood as engineering Markov blankets" is not just a cute rephrasing; it reframes design as engineering the very channels of information and inference. That goes well beyond physical affordances into the realm of probabilistic dependency structures, which most designers never consider.

From a philosophical perspective, the main problem concerns the blanket's status. Based on the previous discussion, I adopt a form of operational realism (Woodward 2003). Operational realism about Markov blankets holds that blankets are neither mere statistical descriptions nor metaphysical essences lurking in the world; they are control-relevant structures in dynamical systems that can be detected, stabilized, and exploited by an observer (designer or user). On this view, a Markov blanket at a given spatial scale s and temporal window t is real when variables designated as "blanket states" sustain robust conditional independencies—formally, they render internal and external states approximately independent—and do so in a way that survives intervention and remeasurement. The key is counterfactual import: if intervening on candidate blanket variables predictably modulates the informational traffic between internal and external states, while interventions elsewhere do not, then the blanket is not just a useful summary but a stable handle on the system's organization. This stance is realist because it anchors blankets in manipulation-resistant invariants, yet operational because it is explicitly scale- and task-relative. It rejects uniqueness without lapsing into arbitrariness by treating blankets as equivalence classes MB (s, t) of coarse-grainings that yield the same interventionist profile; multiple members of the class may exist, but not every partition qualifies. It answers the standard "model-dependence" and "reification" objections by imposing robustness tests—across algorithms for structure discovery, across measurement modalities, and across families of perturbations—such that spurious blankets wash out when the ecology changes, whereas genuine ones remain predictively and causally efficacious. It is stricter than instrumentalism because the warrant is not just predictive fit but reliable control: blankets earn their keep when treating them as real improves planning, fault isolation, and design under distributional shift. It is also falsifiable: if no partition maintains the required independencies under intervention, or if purported blanket interventions fail to produce the expected decouplings or recouplings, the realism claim is withdrawn. Finally, operational realism is methodologically fruitful: it licenses "inversion" procedures in which we engineer or tune candidate blankets and test whether the predicted counterfactuals obtain, thereby turning ontological commitment into experimental discipline. In short, Markov blankets are real to the extent that they are stable, intervention-supporting patterns that afford predictive control at specified scales; that is enough reality for science and engineering, and more than mere bookkeeping.

We can further justify and develop this idea by drawing on the latest advances in active inference research. I refer again to Beck and Ramstead (2025) and Sakthivadivel (2022a), who fundamentally reshaped our understanding of Markov blankets. Beck and Ramstead (2025) develop an unsupervised, dynamic algorithm that detects and classifies statistical boundaries in high-dimensional dynamical systems—far beyond the classic view of a fixed, static border (figure 4.1). Their method demonstrates that a Markov blanket is not an arbitrary modeling convenience but an essential feature that distinguishes macroscopic objects in data. It also tackles the combinatorial explosion of possible blankets—in static settings the number of candidate boundaries grows dramatically as system components multiply (Beck & Ramstead 2025, 9). By embedding a Bayesian attention mechanism that selectively focuses on the most plausible boundary variables, they reduce the search space and build a robust taxonomy of object types based on boundary statistics (for technical details, see Beck & Ramstead 2025, 10–12, and their examples).

We can apply this idea to design. If Beck and Ramstead's (2025) algorithm lets us detect an existing boundary, then by reversing that process a designer can forge a new one. In other words, because Beck and Ramstead show how an agent can infer which elements form a blanket, we can invert their framework to ask: *How might a designer actively forge a new statistical boundary within a complex system?* In Beck and Ramstead's detection algorithm, one alternates between inferring blanket membership—assigning each microscopic element the label "internal," "boundary," or "external" according to which configuration best explains the observed dynamics—and refining the generative model so that those assignments remain self-consistent on a macro level. To reverse this process, a designer embeds focused priors in the generative model that give high probability to a chosen subset of elements as the future blanket. Rather than simply observing, the designer then plans and executes targeted interventions (e.g., rewiring connections, toggling sensors, or rearranging spatial relationships) to drive the system toward configurations that satisfy these priors. Each intervention is selected by evaluating its EFE: actions that promise both epistemic clarity about the emerging boundary and pragmatic progress toward its realization are favored. As the designer's interventions reshape the system, a directed Bayesian attention mechanism continually concentrates computational effort on the candidate boundary variables, refining their assignments in light of new observations. This targeted update accelerates convergence, because it ignores irrelevant components and reinforces the intended separator. Eventually, the system's empirical dynamics come to reflect a clear separation: internal states influence external states only through the newly configured blanket variables. At that point, the statistical boundary has been effectively stabilized—the artifact now behaves as its own autonomous agent, complete with a self-organizing frontier that users can then recognize.

This notion is implicit in Beck and Ramstead (2025), in a key passage: "Given path statistics on a Markov blanket, Jaynes' principle can be used to identify an energy function and associated Lagrangian that specify the dynamics of the environment, boundary, and object variables that give rise to an object of that type" (Beck & Ramstead 2025, 5). They add: "This is consistent with the notion of system typing used in physics, wherein physical systems are defined by the energy functions or stationary actions that they minimize" (Beck & Ramstead 2025, 5). In short, following the path-based formulation of the FEP, to describe the behavior of a system, we need to choose the probability distribution that maximizes entropy under certain constraints (Jaynes' principle of maximum caliber). We can then translate this distribution into a Lagrangian that specifies the cost of deviating from a particular dynamical behavior. This describes which dynamic configurations are more or less probable, given the boundary statistics. Once the Lagrangian is defined, we can derive the equations that describe the evolution of the environment, boundary, and interior of the system. Essentially, if I know how the boundary of an object behaves over time (or how I want it to behave), I can construct (or allow to emerge) that object by specifying the appropriate dynamics according to Jaynes' principle. The objective function that one obtains from combining blanket statistics with Jaynes' principle of maximum caliber "is precisely the free energy" (Beck & Ramstead 2025, 6). In summary, the FEP and the principle of maximum caliber share the same structure. These same principles can be applied to the design of a Markov blanket. We can derive a Lagrangian function that describes what it means to be an object of that type, in dynamical terms. The actions required to construct that boundary depend on this description of the system boundary (i.e., the blueprint). This

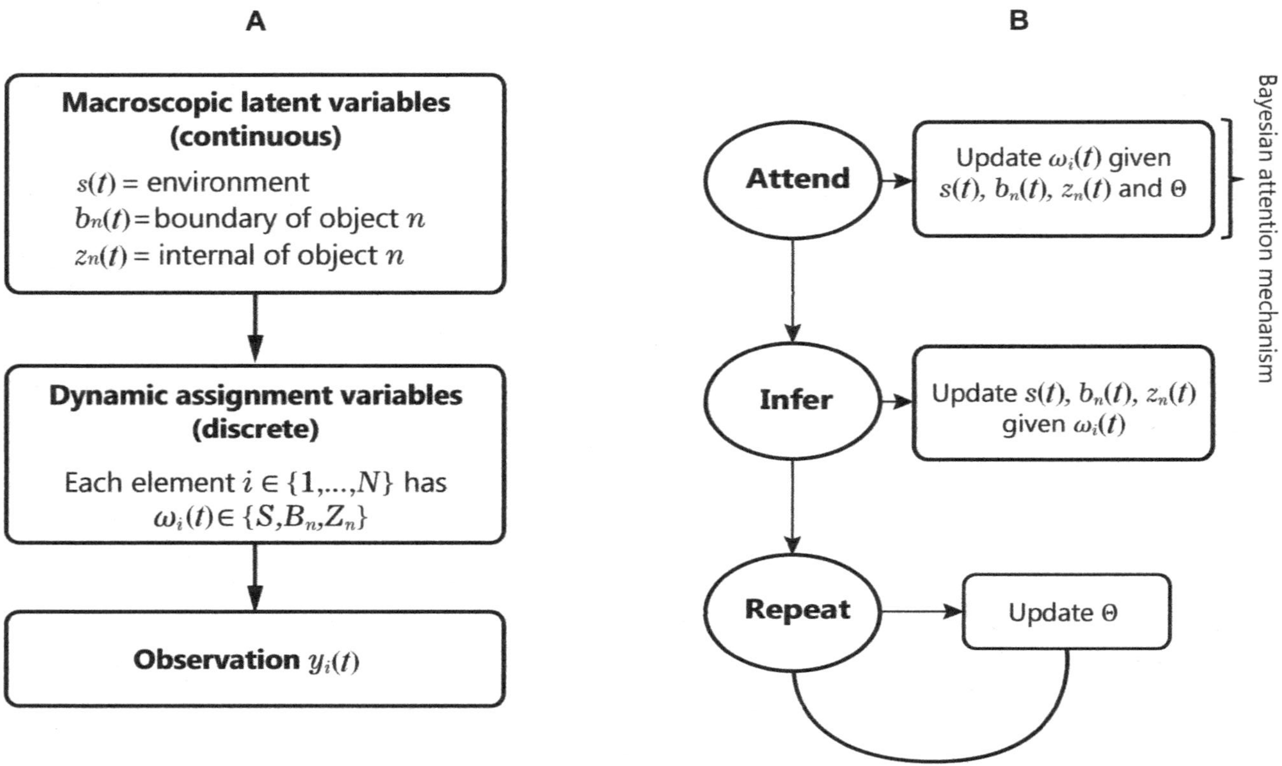
A
Macroscopic latent variables (continuous)
$s(t)$ = environment
$b_n(t)$ = boundary of object n
$z_n(t)$ = internal of object n
Dynamic assignment variables (discrete)
Each element $i \in \{1,\ldots,N\}$ has $\omega_i(t) \in \{S, B_n, Z_n\}$
Observation $y_i(t)$
B
Attend
Update $\omega_i(t)$ given $s(t)$, $b_n(t)$, $z_n(t)$ and Θ
Bayesian attention mechanism
Infer
Update $s(t)$, $b_n(t)$, $z_n(t)$ given $\omega_i(t)$
Repeat
Update Θ

Figure 4.1

A schematic view of the Beck and Ramstead model. Beck and Ramstead (2025) distinguish between two groups of variables: (A) dynamic assignment variables $\omega_i(t)$, and (B) macroscopic latent variables $s(t)$, $b_n(t)$, and $z_n(t)$. The former are discrete, local variables that function as functional labels for each microscopic element or measurement. Each $\omega_i(t)$ can take on one of three values: s (environment), b_n (boundary of object n), or z_n (interior of object n). These variables are called dynamic because they model the transition of a microscopic element i from one role to another, subject to the constraints of the Markov blanket structure—that is, an internal element can interact with an external one only through the blanket, and vice versa. The assignment of these dynamic variables is governed by a Bayesian attention mechanism: the algorithm computes, for each observation (e.g., a pixel or a particle), the probability that it belongs to the environment, an object, or its boundary. This assignment is stochastic and dynamic (time-varying), reflecting changes in the system's perceptual and object-based organization over time. The second group of variables—$s(t)$, $b_n(t)$, and $z_n(t)$—are continuous and macroscopic in nature. They represent the collective state of groups of microscopic elements. For example, all elements labeled as internal contribute to estimating the value of $z_n(t)$. The update algorithm is structured as an iterative process of variational inference and learning, using variational Bayesian expectation maximization (VBEM). This process alternates updates between the dynamic assignment variables, the macroscopic latent variables, and the model parameters Θ (including dynamics, emissions, noise, and other structural components). The algorithm unfolds in three main steps: attend, where the dynamic assignment variables $\omega_i(t)$ are updated based on the current values of $s(t)$, $b_n(t)$, $z_n(t)$, and the model parameters Θ; infer, where the continuous latent variables $s(t)$, $b_n(t)$, and $z_n(t)$ are updated given a current hypothesis about $\omega_i(t)$; and repeat, where the model parameters Θ are revised. The entire process aims to maximize the ELBO, which quantifies how well the approximate posterior accounts for the observed data under the assumed generative model. Therefore, Beck and Ramstead ask: Given a system, does it have a Markov blanket? I ask: Given a desired blanket, how do we engineer a system that converges to it? The formal concept is identical—same definition, same metrics—but the direction is reversed. This is what I mean by inverting the detection algorithm. See https://github.com/DesignAInf/DAU-software-tool.git.

is the foundation for building and refining an artifact's generative models—that is, its fundamental functions. For a more formal explanation of the Beck and Ramstead (2025) model inversion and a computational simulation, see https://github.com/DesignAInf/DAU-software-tool.git.

Does this explain the entire design process? How does the designer establish their priors and transform their generative model? And how does that phase of design—the very heart of the process—known as ideation (and not merely invention) actually occur (Gonçalves & Cash 2021; Lee & Ostwald 2022; Novica et al. 2023)? Obviously, the design process is far more complex and unfolds in several stages. Creating a new Markov blanket requires generating new priors—and thus a new generative model for the designer. This can be understood as a form of structure learning (as noted at the end of the previous chapter). Smith et al. (2020b) define an active-inference–based model of structure learning grounded in the idea that a generative model can be equipped with extra "slots" (hidden states or causes)—that is, undefined priors that can be deployed whenever the agent encounters (or imagines) new observations, concepts, objects, actions. Corresponding to this expansion of the generative model is an inverse reduction phase to ensure efficiency. Structure learning can therefore be seen as an emergent property of active inference: an unsupervised process driven solely by the minimization of variational free energy.

In other words, according to Smith et al. (2020b), to allow new observations, concepts, or actions to emerge, the generative model must expand by acquiring new variables (hidden states) that initially carry no semantic content. These empty slots start with a uniform distribution over all possible observations and remain inert until the agent encounters feature patterns that existing concepts cannot explain. At that point, empty slots can be used to entertain new hypotheses. Each empty slot is gradually shaped by incoming data—co-occurrences of observations and the emerging concept. This expansion phase is followed by a reduction phase: the slots created may be too many, incomplete, or ultimately redundant. To ensure efficiency and prevent conceptual overfitting, a Bayesian model reduction procedure analytically compares the evidence for the full model against possible reduced versions. If the reduction in complexity outweighs the slight drop in predictive accuracy, unused slots are "reset" to their initial state, freeing up cognitive resources for future learning. In this way, Smith et al. (2020b, 3) demonstrate "how particular approaches combining Bayesian model reduction and expansion can reproduce successful concept learning without the need for corrective feedback." (This dual effect of expansion and reduction can also be explained in terms of canalization versus plasticity; see Carhart-Harris et al. 2023.)

More precisely, the "empty slots" refer to potential hidden states for which there is not yet any specific knowledge of how they relate to observations. In a matrix—which encodes the probability of observing each feature given a hidden state—each column represents a possible concept. For empty slots, these columns begin with a uniform distribution (a flat prior), meaning that every observation is initially considered equally likely. In this sense, they are "empty": they contain no distinctive information about any predefined concept, unlike the slots already assigned to "bird," "fish," or other concepts. When the agent encounters a combination of features that cannot be explained by the existing slots, active inference identifies one of these "virgin" slots as providing the best explanation without increasing complexity. At that point, the slot is "filled" by the incoming data: the probabilities in the corresponding column are updated through free-energy minimization, gradually transforming an empty slot into a meaningful one—that is, a newly acquired concept (for technical details, see the simulations in Smith et al. 2020b).

Drawing on Smith et al. (2020b), I propose framing the usual five phases of the design process—research, analysis and strategy, ideation, design, and prototyping and testing—as iterative cycles of expansion and reduction of the designer's generative model, all aimed at forging a new Markov blanket (the artifact) along with its own generative model (the artifact's functions, which depend on the designer's generative model) (see figure 4.2). Consequently, in this case the empty slots apply to the labels—external states, internal states, and blanket states—as well as to the actions needed to actively embed those labels. Research and analysis thus map to the expansion of the existing generative model via empty slots, whereas ideation and design correspond to the reductive phase, in which the designer must decide precisely which product to bring into being. We can then embed the Beck and Ramstead (2025) model into the one proposed by Smith et al. (2020b): each design phase (long timescale) corresponds to multiple processes of defining and redefining the Markov blanket (short timescale). The designer first "opens" new flat-prior slots and probes them via small interventions, letting structure learning (Smith et al. 2020b's model) reveal which labels or actions lower free energy. Once a slot proves explanatory, its posterior is converted into a soft prior over blanket membership (Beck and Ramstead's inverted model), biasing boundary formation toward that new label. Finally, the inverted boundary detector drives real-world interventions to realize the Markov blanket, and the outcomes either confirm the prior or send us back to expand or adjust slots. Once the Markov blanket and generative model are specified, we can model the artifact's behavior in terms of free-energy minimization or co-minimization (i.e., updating beliefs on its environment and the users). The real challenge is that once the artifact leaves the design phase and enters use, the user effectively becomes another designer—able to modify it and develop it in new, unexpected directions. (For this reason, we will later discuss the artifact in terms of co-synchronization.)

Obviously, this is just a conceptual model that will need to be further developed through concrete case studies. Nevertheless, it shows how active inference offers a fresh perspective on design.

Let us try to be clearer. Imagine a meditation pavilion nestled in a Zen garden whose main purpose is not only to provide protection, but also to optimize the meditative experience of its users/inhabitants. The architectural shell should function as a Markov blanket: interior states (interior sound, temperature, illuminance, and a latent "comfort" variable) are conditionally independent of exterior states (weather, ambient noise, solar irradiance) given the boundary's sensory and active states. This means producing and testing a model of the artifact and its Bayes-optimal behavior in a given context.

Exterior variables E drive sensors on the shell S (facade microphones, glazing illuminance probes, envelope thermistors). Interior variables I drive actuators A (louver angle, glazing opacity, roof tilt/ventilation). The design target is the factorization

$$p(E, S, A, I) = p(E)\, p(S|E)\, p(I|S, A)\, p(A|I)$$

which encodes the blanket relations. Here S denotes environment-facing sensors placed so that their readings are not directly affected by actuators; any actuator-induced boundary effects are modeled as separate variables. Sensors are environment-facing transducers: they change with E and do not depend directly on the interior. Actuators are governed by interior error signals relative to designer priors (comfort bands); raw exterior variables never set actuator states directly. Once S and A are fixed, alterations in weather or street noise cannot alter interior fields except through those channels. That engineered shielding—$I \perp E \mid (S, A)$—is what turns the

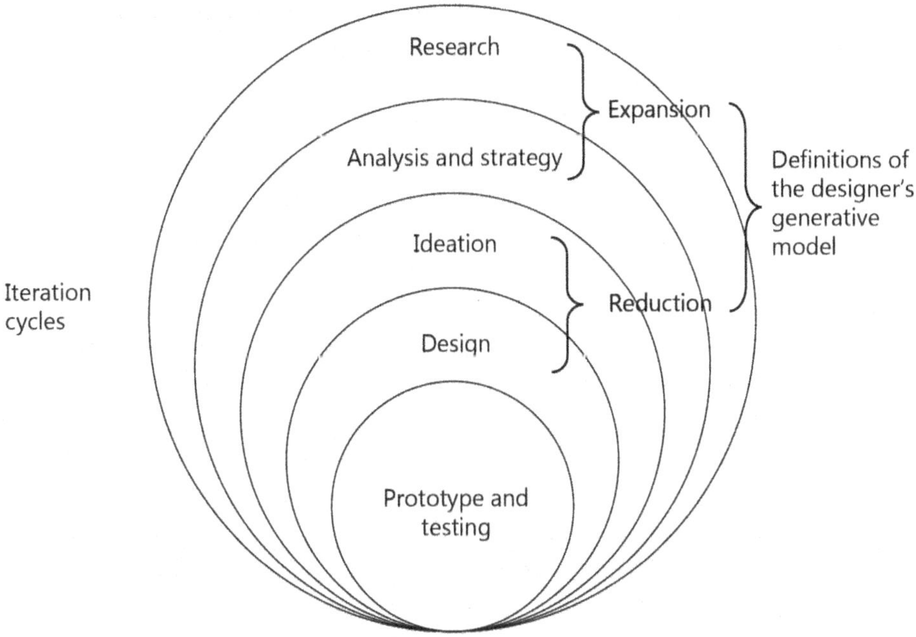

Figure 4.2
The five phases of design reinterpreted through the concept of structure learning in active inference and the inversion of the Beck and Ramstead's model. The designer first dives into research to uncover user needs and problems. They then analyze findings to craft a smart strategy. Next comes ideation, a burst of creative brainstorming. This flows into design, shaping refined solutions. Finally, prototypes are built and rigorously tested to perfect the outcome. These phases mirror established models like the Double Diamond or IDEO's human-centered design process.

pavilion into a blanket. As design constraints (and to a first approximation), we impose three key screening-off relations:

(i) $I \perp E \mid (S, A)$: the internal and external become independent once the sensors and actuators are known; this is the operational definition of a Markov blanket.
(ii) $S \perp (I, A) \mid E$: the sensors, given the external, do not directly depend on the internal state or on the actions (no "remote control" of the input).
(iii) $A \perp (E, S) \mid I$: actions are not set directly by external variables; sensory inputs influence action only via interior (inferred) states (avoiding shortcut control).

Can we turn this into an actionable blueprint? And what are the advantages of doing so? Start by specifying comfort priors (e.g., 35–45 dB, 21–23°C, 200–400 lux). Use materials and geometries to shape precision at the boundary—acoustic absorption and panel stiffness to down-weight nuisance vibrations; electrochromic glazing to tune illuminance likelihoods; thermal mass and controlled ventilation to stabilize heat exchange. Add curiosity without chaos: low-variance cues (water sound, slow air movement) that afford epistemic value but do not destabilize the boundary. Embed predictions in policy by making $A \leftarrow I$: control laws act on interior deviations from priors rather than on exterior readings, so adaptation occurs from the inside out. Instead of "detecting" a preexisting blanket, the designer creates one by iterating structure. Begin broad—prototype sensor/actuator "slots" and boundary elements that might realize the factorization—then prune whatever fails to improve the intended conditional independencies or violates the mapping $S \leftarrow E$, $A \leftarrow I$. Lock in policies so actuators respond rapidly to interior errors while sensors read the exterior with minimal back-effect. The cycle is expansion (try slots) → reduction (remove passengers) → tuning (tighten priors and gains).

From a FEP perspective, the pavilion's generative model is not a symbolic representation inside the artifact, nor is it the designer's model. It is implicitly embodied in the pavilion's material organization and control architecture. The Markov blanket supplies boundary conditions: sensors implement mappings from external states to sensory signals, actuators shape the sensory consequences of action, and comfort bands encode prior preferences while precision settings weight which deviations matter most. In this sense, the pavilion is an agent: it preserves a viable regime by acting to keep exchanges across the blanket within expected bounds (self-evidencing). This also yields a built-in notion of sustainability as viability: preferences delineate an operating envelope, and self-evidencing consists in maintaining the system within that envelope under perturbation. Ecological performance (e.g., energy and material constraints) and user well-being can be included to the extent that they are made explicit in those preferences and constraints. Psychological sustainability can be modeled as preferences over sensory regimes that reduce cognitive load—low volatility in noise and glare, smooth transitions in illuminance, and predictable microclimate dynamics. By keeping sensory exchanges within these expected bounds, the pavilion supports attentional stability and sustained comfort, rather than repeated episodes of sensory surprise.

In a nutshell, preferences (comfort bands) directly define the viability set; the system acts to remain within it by reducing risk and uncertainty. The pavilion does not merely react to spikes in noise or heat, but anticipates, dampens, and compensates in an informed way. Sustainability here is not "eco-friendly" in some vague sense, but *inferential stability* over time.

Validation can be done empirically on site. First, vary exterior conditions (pink-noise outside the shell; controlled shading; warm-air blasts) and confirm that sensor readings track those changes while the interior remains in band because actuators compensate—evidence for $S \leftarrow E$ and an effective $A \leftarrow I$ loop. Second, perturb the interior directly (brief interior sound/heat/light injections) and verify that actuators respond appropriately with no exterior shortcut to A. Third, hold S and A fixed (replay recorded sensor traces; freeze actuator setpoints), then sweep exterior conditions and check that interior fields no longer vary systematically—an operational test of $I \perp E \mid S, A$. Finally, audit for leaks with smoke, or laser vibrometry at seams; any residual direct $E \rightarrow I$ paths are either sealed or explicitly acknowledged as local E' nodes at the shell. Report the fraction of time the interior remains within comfort bands under low/high exterior set-points for each modality, along with a brief note on any conditional independence (CI) test that fails and the mitigation taken. This is sufficient evidence that, within its design envelope, the pavilion boundary functions as a Markov blanket. Seasonal growth or site changes may drift the $S \leftarrow E$ mapping, requiring retuning. Open doors or occupant overrides create temporary non-mediated paths; treat these as policy exceptions rather than blanket failure. Actuator exhausts can locally perturb the exterior; model these as adjacent E' nodes to preserve the CI statement.

The designer does more than simply assemble sensors and actuators; they explicitly define the conditional dependencies among internal, external, and blanket variables. On one hand, this approach makes the effectiveness of the design quantifiable—the degree to which the product (i.e., the blanket) protects and guides the system can be measured in terms of free energy and predictive accuracy—and on the other hand, it paves the way for automated optimization tools based not only on a data environment but also on physics. In other words, under this approach the designer *engineers an inferential boundary*—and the system's ability to maintain that boundary. There is a single objective: minimizing free energy, that is, co-minimizing free energy together with the user, rather than juggling multiple objectives as in classical design (e.g., satisfaction, conversion—when the user completes the desired action—time-on-task, Net Promoter Score [NPS]). Design as engineering Markov blankets aims to reduce costly surprises for relevant agents (users, operators, stakeholders) given a set of stated expectations. Free energy minimization is the integrative dashboard that helps combine security, usability, costs, impact, and user experience.

Therefore, design on the account developed here, is not adequately characterized as the production of objects, interfaces, or affordances in isolation, but as the engineering of inferential boundaries that mediate and stabilize interactions between systems. Within the active inference framework, such boundaries can be formally described in terms of Markov blankets. Importantly, however, not every designed artifact constitutes a Markov blanket. The concept applies only when design stabilizes inferential boundaries that meet minimal criteria of functional screening-off, bidirectional sensor–actuator coupling, and robustness under intervention. In this sense, Markov blankets are not metaphysical entities nor merely post hoc modeling conveniences, but operational constructs whose relevance depends on their persistence across perturbations and their capacity to regulate exchanges between systems. Design, understood in this way, can be seen as the deliberate construction and modulation of such operationally real inferential boundaries, thereby enabling new forms of coordinated behavior, control, and agency within sociotechnical systems.

There is a crucial objection that needs to be addressed directly. Any engineer working on active inference would say: "We already design and build active inference agents in our daily work—when we design the generative model of an agent, we are already implicitly designing its Markov blanket. So what is the point of talking about 'inversion' or 'designing Markov blankets'?" This is an important objection, but it misses a key point.

Design does something fundamentally different from engineering practice. First, the designer is concerned with the relational dimension—the relationship between what is being designed and the human being who will use it. This necessarily means including the user's psychological world in the process; the designer does not work in a vacuum. Second, the designer's primary focus is form. Rooted in the phenomenological tradition from Husserl to Merleau-Ponty and Vial (2017), form is not a property of the thing itself—it is what structures the perceptual and motor experience of an agent in the encounter between the artifact and the body that perceives and uses it. A door handle has the shape it has because a human hand grasps it in a particular way. Form is relational. More precisely, form is about the production and structuring of affordances.

If we accept these two premises, the fundamental question becomes: given a set of data and constraints, what is the best form for what we are trying to build? This is precisely where "design as creating and shaping Markov blankets" becomes useful. The notion of the Markov blanket allows us to conceptualize form in a new and more rigorous way: *as the set of conditional independencies that define the sustainability of the designed object.* Engineering applications in active inference typically take as a precondition the regime of interaction within which a system must operate; design treats that regime as the primary problem. The designer does not start from the generative model to be implemented but from the specification of the informational boundary that must make possible a certain relationship among user, artifact, and environment. Crucially, the Markov blanket does not determine a single generative model—it defines the space of models compatible with that designed relationship. The designer's work therefore cannot be reduced to making a generative model function: it concerns the very definition of the agent-environment relationship the project must bring into being.

4.2 THREE DESIGN MECHANISMS: PRECISION CRAFTING, CURIOSITY SCULPTING, AND PREDICTION EMBEDDING

I identify here three mechanisms that the designers use to plan interventions: precision crafting, curiosity sculpting, and prediction embedding. Rather than strict phases, they function as complementary levers the designer can pull—sometimes in parallel, sometimes in sequence—to define a new statistical boundary. Through successive cycles of broadened exploration, focused attention, and durable expectation-setting, the designer incrementally defines the very statistical separations that define a new autonomous entity. There is, then, a seamless continuity between Beck and Ramstead's abstract framework and the phenomenological design structures outlined here. Each of the three intervention levers (precision, curiosity, prediction) corresponds to a key stage in the protocol for engineering Markov blankets.

Precision crafting refers to how designers can shape precision-weighting. For example, road signs in a traffic system direct a driver's attention to critical information such as speed limits or directions. In this way, the artifact structures sensory input and helps the agent focus on more relevant prediction errors.

Graphical design elements play a crucial role in shaping precision, much like language influences the weighting of specific prediction errors (Clark 2016, 283). For instance, lines effectively guide attention: straight lines convey order and stability, while curved or diagonal lines introduce dynamism and encourage exploration. Shapes, too, carry distinct affordances—circles evoke inclusiveness, while squares and rectangles suggest stability and functionality. Color enhances precision by signaling salience. Warm tones like red draw focus and convey urgency, whereas cooler hues like blue promote calm and sustained engagement. High-contrast combinations, such as bright yellow against a dark background, further emphasize critical elements, reducing ambiguity and sharpening perception. Texture adds another layer of influence by shaping user interaction. A rough surface signals caution or grip, while smooth textures invite ease of use. Additionally, space, scale, and proportion organize visual relationships: negative space highlights key elements, proportion maintains harmony, and scale directs attention—such as oversized buttons that encourage immediate interaction. By integrating these elements, designers can shape and refine users' generative models, reduce cognitive load, and align expectations with an artifact's intended function—thus effectively narrowing the environment. In this way, graphical design serves as "a finely tuned means of artificially manipulating the precision (hence temporarily modifying the impact) of prediction error at different levels of neural processing" (Clark 2016, 284). Consider the iconic 1931 London Underground map designed by Harry Beck, which revolutionized urban navigation. By shifting the focus from geographic accuracy to a topological representation, Beck's design fundamentally altered how users understood and predicted navigation. The map not only improved the ease of route planning but also had profound cultural and behavioral impacts, shaping the way people interacted with and conceptualized the city. The map can be seen as a precision crafting device. It serves as a nonhuman agent, reshaping a network of interactions between humans (London passengers) and nonhumans (the subway system and the city itself). The map functions as a generative model that guides action and continuously updates the recognition model encoded in passengers' brains. Part of the computational process occurs through nonhuman agents—in this case, the map.

Curiosity sculpting is the process where designers can create forms of controlled uncertainty in the environment, driving the pursuit of knowledge. In this process, the designers seek out or create spaces of uncertainty, that is, "limited forms of simulation" (Clark 2016, 275; see also Constant et al. 2024), that challenge their current predictions, thereby promoting learning and exploration. Here again, the material properties of the artifact play an active role. Scientific instruments, such as particle accelerators, create controlled environments where uncertainty is maximized, generating unpredictable data that scientists can use to refine their predictions about the physical world. The material design of the accelerator itself sculpts curiosity by introducing uncertainty that is difficult for the agent to predict directly, making it possible to push the boundaries of existing knowledge. On a more everyday level, toys like video games, puzzles, or building blocks create environments of controlled uncertainty for children. These material objects prompt children to experiment, test, and revise their predictions, fostering curiosity—and producing delight in them. In this case, curiosity sculpting provides opportunities for the agent to encounter new information that was previously unpredictable, driving cognitive development.

It is important to recall a crucial distinction: not all uncertainty possesses epistemic value. Undesirable uncertainty refers to unpredictability that offers no useful information for refining the generative model and may even disrupt an agent's ability to function effectively within its environment. This kind of uncertainty introduces noise or randomness that overwhelms the agent's predictive capacities without contributing to better future predictions. For example, consider an autonomous vehicle experiencing sensory glitches from malfunctioning sensors. The vehicle's system might register random, irrelevant data—such as falsely detecting obstacles—that do not reflect the true state of its environment. This undesirable uncertainty misleads the human driver, causing faulty predictions and potentially erratic decisions, without aiding in the refinement of its generative model. Active inference agents, therefore, need mechanisms to filter uncertainty based on its epistemic value. Desirable uncertainty—uncertainty that holds potential for improving the generative models and the inferential process—must be distinguished from irrelevant or disruptive uncertainty (Parr & Friston 2017; Parr et al. 2022, 217–218).

Prediction embedding involves the externalization of predictions, meaning that artifacts encode or embed a set of expectations about the environment and human behavior. In this case as well, every externalization of the user's generative model corresponds to an internalization—that is, an update of the user's recognition model. And vice versa: every internalization depends on how the generative model is translated into actions and artifacts. It is not simply an extension of cognitive faculties that occurs.

Baltieri et al. (2020) demonstrate concretely how an artifact embeds a generative model. Moreover, artifacts themselves can also generate predictions. Large language models (LLMs), for example, are designed to encode and replicate intricate networks of linguistic and cognitive expectations. Through extensive training on vast datasets, an LLM develops the ability to anticipate the next element in a sequence of tokens based on a given prompt from a human user. By leveraging these learned expectations, the model generates responses that are, in most cases, contextually relevant and coherent, thereby aligning with the inquiry at hand. This predictive capacity enables the human user to reduce uncertainty by providing structured and informed suggestions. In this dynamic, the LLM does not merely react but actively guides interaction by offering insights and potential solutions—AI can encode users' generative models, or parts of them. Nonetheless, Di Paolo et al. (2024) rightly point out that LLMs often rely on a form of passive prediction and learning.

Even more strikingly, internet recommendation systems and data mining techniques can exhibit more refined and active forms of prediction. Unlike LLMs, which typically rely on explicit prompts and offline training, recommendation algorithms operate proactively, continuously anticipating user preferences and behaviors in real time. These systems adjust their predictions based on user feedback, shaping choices before they are even consciously made. In doing so, they exemplify an even more autonomous form of machine-driven prediction, influencing human decision-making in subtle yet profound ways.

Now, Constant et al. (2021) validate all these concepts regarding the three mechanisms described above. Their study, using active inference-based simulations, demonstrates the acquisition of culturally patterned attention styles through artifacts. The authors show that increased complexity of decorations on ancient pottery directly influences cognition—particularly visual perception—producing "culturally patterned attention styles" that are "the result of the agent attempting to predict the next decoration based on observed pigments and learning the probability transitions between the visual motifs forming the decorations" (Constant et al. 2021, 2). The paper builds on the work of archaeologists (Criado-Boado 2014; Criado-Boado et al. 2019) who have examined the relationship between the evolution of pottery decoration and shifts in social complexity—exploring how artifacts such as architecture, metallurgy, jewelry, media, monuments, and tattoos reflect a society's organizational structure and the expression of social identities. The simulations provide a compelling proof of concept of a link between interactions with artifacts and the acquisition and development of "attention styles based on the exposure to vase decoration" (Constant et al. 2021, 3). For example, the first experiment demonstrates that variations in decorative complexity influence the agent's gaze patterns as the vase is explored, affecting attention levels and the learning process. The agent's gaze adopts an attention style and, consequently, a generative model shared with other agents—further corroborating Veissière et al. (2020). This is a crucial insight: the vases' decorations embody certain expectations (specific gaze patterns) and actively shape attention.

In more technical terms, Constant et al. (2021) assess the impact of decoration patterns on visual saccades using the Vertical Index (*Vi*), a metric that quantifies the relative prevalence of vertical versus horizontal saccades during visual exploration. They employ an active inference model to simulate saccadic behavior while artificial agents examine decorated antique vases. The *Vi* is computed as the difference between vertically and horizontally directed saccades, weighted by the display's height and width, respectively. The results indicate that increasing the complexity of vertical decorative patterns leads to a higher *Vi*, meaning a greater proportion of vertical saccades. This suggests that material culture, specifically the arrangement of visual motifs, influences perceptual strategies by shaping predictive saccadic behaviors.

Another example in design, indoor lighting, is especially important—and it's also connected to the example of the pavilion in the previous section. We spend about 90 percent of our time in artificial environments, and lighting profoundly shapes our cognitive abilities and emotions. Furthermore, lighting has profound effects not only on vision and cognition but also on the biological and psychological conditions of human life (Boyce 2021). In shared spaces, where different agents perform different tasks and have different preferences, achieving a shared lighting setup that adequately meets everyone's needs can be very challenging. This becomes even more complex if we consider the different ways light is measured and evaluated: lumens (the brightness of the light source, i.e., how much total visible light a lamp emits), lux (the amount of light that actually reaches a surface, depending on distance and direction), and kelvin (the color appearance of the light: low Kelvin = warm, yellowish; high Kelvin = cool, bluish) (Castilla et al. 2023).

From an active inference perspective, the crucial question is: How does lighting influence the trade-off between exploration and exploitation—between information gain and pragmatic value—in the behavior of each agent and of the group as a whole? In design terms: How can we model the behavior of the artifact (i.e., the shared lighting system) so that it supports a good trade-off? Lighting design can be understood as a regulator of the agent's sensory precision.

This problem becomes very concrete if we imagine two people reading at the same desk: one is studying, the other is reading a novel. Their tasks and preferences differ, yet they share the same environment and the same lighting conditions. How do we find a common lighting solution that allows each person's eye movements to follow a direction and speed appropriate to their respective goals? In this specific case, it is plausible to assume that, for a reading agent, a high level of lux may reduce uncertainty in word perception and allow a more efficient and stable strategy (e.g., rereading a passage, slowing down, making more complex inferences, consulting context, activating deeper semantic memory or knowledge, mentally reformulating what has been read). Conversely, periodic variation in illuminance from the light source could increase uncertainty and therefore promote exploration over exploitation. In other words, if lighting is more variable, fixation duration and saccadic patterns tend to change. Light modulates which parts of the visual field are treated as most useful for reducing uncertainty, and thus directly influences saccades. This also means we can deliberately modulate light—using multiple light sources and colors—to optimize saccadic behavior in relation to the task, the agent's condition (e.g., fatigue), and the environment.

4.2.1 HOW CAN ARTIFACTS ENCODE PROBABILITY DISTRIBUTIONS?

If we embrace a symmetrical ontology grounded in the concept of Markov blanket, it becomes entirely plausible to view an artifact as encoding a generative model of the form $P(A|D)$ and, by extension, $P(O|A, D)$ —where A denotes action, D denotes design choices (i.e., the model), and O denotes outcomes. A messaging app, for instance, may also be understood as a recognition model that maps the relation between sensory inputs and external states, and which the user (here the generative model) continuously updates through interaction. In this case, we have an inferential process distributed between two agents: one plays the role of recognition model (maps the relationship between observations and hidden states) and another plays the role of generative model used to update the recognition model (Baltieri et al. 2020).

Table 4.1 provides a synthetic view about how an artifact's material constitution can encode probability distributions (Durrett 2019). (Most of the literature on active inference assumes Gaussian models, for a variety of practical and theoretical reasons. However, this does not mean that other distributions are incompatible.)

The key insight driving this analysis is that artifacts are not just passive objects but probabilistically structured entities that shape and constrain interactions at multiple levels. They can encode expectations about forces, sensory feedback, and structural adaptation within their very materiality. The first way an artifact encodes a probability distribution is through geometric constraints on action. An artifact's shape and mechanical affordances define a probability distribution over interaction forces and motion paths. By physically restricting certain movements while allowing others, artifacts such as door handles, screwdrivers, and machine interfaces can encode Gaussian and beta-distributed expectations over applied force and directionality. The second way, surface texture encoding sensory expectations, focuses on how tactile interaction is probabilistically regulated by friction, resistance, and adhesion properties. Surface microstructure defines the distribution of grip stability and touch dynamics, shaping user expectations through Gaussian and Dirichlet distributions over friction coefficients and anisotropic contact forces. A touchscreen or rubberized grip stabilizes interaction variability by encoding a well-defined probability range over expected resistance. The third way, internal material structure encoding probabilistic responses, describes how an artifact's molecular and mechanical properties determine its probabilistic state transitions under stress and environmental forces. These transitions can be modeled using Markov matrices, Weibull distributions for material fatigue, and gamma distributions for structural adaptation. Examples include shape-memory alloys that revert probabilistically to their original form or polymers that exhibit entropy-minimizing deformation responses.

Table 4.1

A quantitative, material-based perspective on how artifacts constrain, guide, and regulate interaction through probabilistic means.

Types of Encoding	Explanation	Types of Probability Distribution Encoded	Examples
Geometric constraints on action	The artifact's geometry imposes structured constraints on movement and force application, defining a prior probability distribution over user interactions.	Multivariate Gaussian distribution over interaction forces (defining expected mean force and variance); Beta distribution encoding preferred vs. restricted motion paths	A door handle enforces a probability distribution over pulling vs. pushing actions; a screwdriver's shaft constrains torque application to expected rotational ranges.
Surface texture encoding sensory expectations	The microtexture and material properties regulate tactile resistance and frictional forces, shaping a probabilistic mapping between applied force and sensory feedback.	Gaussian and log-normal distributions over frictional forces (modeling grip stability); Dirichlet distribution encoding anisotropic effects of textured surfaces on multi-directional force application	A touchscreen with a polished surface exhibits a low-variance frictional profile for stable haptic feedback; a rubberized grip reduces unpredictability in grasp forces by stabilizing resistance.
Internal material structure encoding probabilistic responses	The molecular and mechanical structure of an artifact governs probabilistic responses to stress, strain, and external perturbations, defining adaptive transitions between material states.	Markov transition matrix defining probabilistic material state changes; Weibull distribution governing fatigue failure probabilities; gamma distribution encoding adaptive structural responses to external stress	Shape-memory alloys exhibit probabilistic state transitions, reverting to their original configuration when thermal thresholds are exceeded; viscoelastic polymers adjust deformation entropy-minimizing recovery patterns, probabilistically optimizing structural resilience.

4.3 ARTIFACTS AS SYNCHRONIZERS

How does an artifact's agency co-minimize free energy alongside that of the designer and the user? My claim is that what truly defines an artifact—and sets it apart from a natural object—is that (1) its statistical boundaries and generative models are created by a designer, and (2) it synchronizes the generative models of multiple agents (designer and users). The function we ascribe to an artifact is precisely the synchronization of the designers' and users' generative models through the artifact's own generative model. This does not rule out autonomous agency in artifacts. Indeed, agency is distinct from function, and precisely this distinction makes possible what postphenomenologists call multistability, that is, the simple fact that the function or use of an artifact can change from one social context to another (Rosenberg & Verbeek 2015); the distinction between agency and function is therefore a conceptual move of great importance. A designer (or team of designers) creates an artifact that, once brought into being, becomes an autonomous agent capable of reshaping the world. A user can only use the artifact correctly by synchronizing with the artifact's Markov blanket and generative model—and therefore indirectly with the designer's generative model. The key point is that designers and users can synchronize only through the artifact itself. Once we adopt this triadic communication structure, the simple metaphor of extension no longer holds. From this perspective, the mind does not simply extend; it synchronizes with other minds through artifacts.

4.3.1 REMARKS ON SYNCHRONIZATION

I assume that *every* artifact has someone who designed and created it so that someone (themselves or someone else) can use it for some purpose, some function. In an ideal scenario, the role of the artifact is enabling synchronization, that is, interactive alignment, between the designer and the user(s). The emergence of such synchronization indicates that the artifact is being used for the purpose for which it was created—designer and user understand each other in the sense that they can easily predict each other's beliefs. This does not preclude the user from becoming a designer and creating new features and functions. By enabling synchronization between agents (designer and user), the artifact supports joint free energy minimization.

One might object that, even if we grant that artifact–user synchronization is a necessary condition for technology, it makes no sense to talk about synchronization between designer and user. After all, when we use our smartphones, we do not ask ourselves, Why did the designer make this choice rather than that one? We just use the device and try to predict how it will behave step by step. But my point is different: artifact–user synchronization implies designer–user synchronization, because the artifact embodies and reflects the designer's generative model. In other words, synchronization is not so much about asking how to use an artifact in a specific situation; it is more a structural condition of technological experience. When a user engages with the artifact, they are effectively interacting with a delayed echo of the designer's predictions. The real distinction lies in timescale: designer–user synchronization is slower and mediated through the artifact (and its earlier and later versions), whereas artifact–user synchronization is faster and more immediate.

I therefore introduce the notion of deferred synchronization to describe a specific mode of predictive alignment between agents who are not co-present in time or space, but who interact with a shared technological artifact (for an overview of the concept of synchrony, see Pikovsky et al. 2001). In this framework, the artifact serves as an embodied repository of expectations, enabling deferred coordination between designer and user. Synchronization occurs not through direct interaction but through a distributed predictive loop over multiple timescales. Deferred synchronization extends the logic of Markov blankets beyond real-time coupling, embedding predictive models within sociotechnical ecologies that persist across temporal and spatial gaps. Synchronization among the designer, the artifact, and the user thus takes place across multiple timescales and at various levels of complexity.

Let us delve into this more. As mentioned earlier, Friston and Frith (2015b) introduce a model of generalized synchrony to explain how interacting active inference agents communicate through mutual prediction. Their approach posits that agents do not explicitly predict each other's actions; rather, each agent predicts its own sensory states, which are influenced by the actions of another agent. Synchronization emerges when these predictions become reciprocally constrained. Put simply, "generalized synchronization means that knowing the state of one system (e.g., neural activity in the brain) means that one can predict the states of the other (e.g., another's brain)" (Friston & Frith 2015b, 391). In other words, if we have (1) two active inference agents, (2) each trying to predict the other, and (3) both operating with similar generative models, then we see generalized synchrony emerging. As said, this means that knowing the state of one agent allows us to predict the state of the other.

In the birdsong model proposed by Friston and Frith (2015b), two birds do not merely react to each other but, over time, develop a form of mutual predictability in their vocalizations, that is, "*identical synchronization* in which there is a one-to-one map between the states of the two systems" (Friston & Frith 2015b, 391). This dynamic entrainment is based on sensory attenuation: "When one bird was singing the other was listening" (Friston & Frith 2015b, 399). This entails the emergence of a structured, reciprocal dependency, where the internal states of both birds become coupled in a way that allows for implicit communication. This reflects a deeper level of interactive alignment beyond mere self-predictability. If two active inference agents attempt to predict both their own behavior and each other's—based on the sensory data they generate—it follows logically, as dictated by the FEP, that their internal states will inevitably converge on the same generative model. And this is mirrored in external states too.

Now, can we apply this concept of generalized synchrony described by Friston and Frith (2015b) to the interaction between designer, artifact, and user? This interaction is conceptually different from that of birdsong. The Friston and Frith model has three essential characteristics: (1) birds can hear each other directly, (2) they share similar generative models, which is connected to the fact that (3) they converge toward a common attractor manifold—a stable set of states. In contrast, designers, artifacts, and users do not interact directly—at least not necessarily. Designers and users rarely engage with each other directly (except in some phases of the design process), while users and artifacts always interact directly. However, the crucial point lies elsewhere: designers and users do not share similar generative models. The artifact embodies the generative model of the designer who created it, while simultaneously integrating the user's model through its practical use. Over time, the artifact itself gradually becomes an attractor manifold toward which both designer and user gravitate. For instance, the designer of the piano embeds a Markov blanket and a generative model in its construction: the layout of the keys, the tuning system, the mechanical action, and the range of possible sounds. These design choices encode assumptions about how music should be played and structured. The user (a pianist) brings their own generative model, shaped by their skill, cultural background, and personal expression. They interact with the piano according to their intentions, exploring its possibilities and adapting to its affordances. Over time, the piano itself becomes a shared generative model: pianists learn to play within the constraints and possibilities of the instrument, while future piano designers refine their designs based on how musicians use them. The piano itself gets in tune with the musician, adapting to the way the musician plays (i.e., the movement of the hands, the body, the rhythm of the breath) and interprets the music. The piano is not just a passive medium but an evolving attractor manifold that aligns both designer and user toward common conventions, techniques, and expectations—and may even alter them.

The example of the piano is particularly interesting because, in this case, the designer and the user have completely different generative models, even though both are focused on the same artifact. The point is that the designer does not prescribe the "correct way" to play the piano being built. Rather, the designer defines the optimal affordances—those that should place the user in the best possible conditions to express their own style and way of playing. The designer's generative model is highly abstract and operates over a much longer temporal horizon than those of the user and the artifact. This leaves significant room for user creativity. The user, in turn, focuses on playing—that is, on making the best use of those affordances to realize their interpretation of the piece, not of the piano. In short, the designer's generative model concerns the interaction between the expected observable consequences of certain design choices and the latent causes that generate them, whereas the pianist's generative model concerns the interaction between the hidden states relevant to performance—e.g., phrasing, harmonic structure, desired timing—and policies/actions that achieve those intended sensory consequences. In this case, the coordination between generative models concerns the definition and use of common expectations and regimes of attention. The alignment of generative models is grounded in a deeper alignment of Markov blankets—the system's statistical and inferential architecture.

This is also what happens in narratives (Clark 2016, 286). When an agent reads a novel, they start with little to no knowledge of what lies ahead—though they inevitably carry priors shaped by previously read novels. Consequently, their generative model differs significantly from that of the novel's author—although there must be some similarities, as a minimal level of communication is essential. What happens next? The novel serves as an active mediator: its genre, style, structure, language, and even its material aspects (e.g., the cover) enable the reader to gradually reconstruct a generative model capable of following the story (Kukkonen 2020, 2024; see also Bouizegarene et al. 2024). The designer—in this case, the writer—develops a set of predictions about the effects of the story and its potential impact on the reader. At the same time, as the reader engages with the text, they generate their own expectations about how the narrative will unfold. The artifact—the story itself—is a materialization of the writer's predictions, crafted to shape the reader's interpretation. However, it is also influenced by the reader's expectations, as their interpretation reconstructs and reshapes its meaning, making the story an evolving entity rather than a fixed message. The reader's generative model can come very close—though never entirely identical—to that of the author. This difference, however, is not a fault or a disadvantage: it enriches the novel and opens up a space for new interpretations and developments—forms of good, controlled uncertainty.

These examples allow me to underline a crucial point. The synchronization of the three agents (designer, artifact, and user) does not imply that one artifact function is "better" than another or that one design outperforms another. Rather, it formalizes the human–artifact relationship in terms of the FEP. The interaction among designer, artifact, and user is not a simple synchronization. In fact, while any multi–active inference network can synchronize when models are similar and coupling is strong, this three-way design cycle—featuring heterogeneous models, partially divergent goals, and artifact-mediated communication—demonstrates how inferential alignment can emerge naturally without any external coordinator.

4.3.2 INNOVATION AND GROUPS

In the first part of this chapter, we focused on design as invention. The challenge, however, lies in defining what makes an artifact "new." An artifact's novelty (and even its acceptability) is always relative to the social group in which it emerges. My thesis in the second part of this chapter is that an artifact's novelty and acceptability are determined within specific domains of the group—what I call "areas of technological innovation." To support this claim, I will focus particularly on the works of Heins et al. (2024), Manrique et al. (2024), and Walker (2025).

First, however, we need to answer a fundamental question: How do active inference agents interact? Let us assume a group of active inference agents, each equipped with a limited number of generative models. Following Pezzulo et al. (2021), generative models can differ along several dimensions:

- the number and type of sensory modalities involved
- temporal depth
- hierarchical depth
- precision of priors
- evolutionary plasticity (the capacity for priors to change).

Moreover, according to Pezzulo et al. (2021), different generative models reflect adaptation to distinct ecological niches and are shaped both by evolutionary inheritance and by the complexity of the niche itself. More sophisticated generative models do not replace simpler ones; instead, they layer on top of them, forming hierarchical structures that can be selectively activated depending on the context. Ecological and social complexity demand greater temporal and hierarchical depth, high prior precision, and adaptive plasticity. Thus, in particularly complex niches that require high levels of foraging, some agents in a group may develop more advanced generative models—and, through synchronization, may influence the rest of the group. However, in large groups, synchronization becomes more difficult. In this case, direct synchronization, which is based on directly receiving the sensory data generated by another agent (as in Friston and Frith's model), is inherently limited. As a result, the spread of highly advanced generative models is likely restricted to smaller subgroups, which must develop more refined, indirect forms of synchronization to understand and influence (and command over) others.

Manrique et al. (2024) apply active inference and the FEP to the study of behavioral evolution in the genus *Homo* during the Pleistocene, presenting it as an "economical, non-trivial, and coherent" normative approach (Manrique et al. 2024, 41). Specifically, they use these frameworks to analyze the case of the Paleolithic hand axe and its irregular, discontinuous appearance over a span of about 2 million years—hand axes first appeared around 1.76 million years ago and remained in use until roughly 100,000 years ago. Traditional explanations rely solely on cultural transmission accumulation. However, according to Manrique et al. (2024), this approach is insufficient to explain how such objects emerged independently across widely separated geographical regions and over hundreds of thousands of years. Instead, the FEP can be viewed as "providing a basis for a normative systematic explanation of the evolution of cognitive abilities in the genus Homo that is compatible with a Paleolithic record of regular irregularities and irregular regularities (e.g., appearances, disappearances, and reappearances in it of 'hand axes')" (Manrique et al. 2024, 42). In other words, these irregularities are "explicable as an emergent property of various free energy minimizing processes that have impinged, at ontogenetic and phylogenetic scales, on the evolution of our behavior" (Manrique et al. 2024, 42). Crucially, according to Manrique et al. (2024), the FEP fundamentally alters our conventional view that human cognitive capacities evolve in a linear fashion. Early members of our genus *Homo* did not advance toward ever-greater technical skill in a smooth, ladder-like progression. Instead, their behavior and its archaeological traces follow a "snakes and ladders" pattern of sporadic random innovations that appear, vanish, and reappear over hundreds of thousands of years.

Recognizing the affordances offered by the hand axe was not immediate. According to Manrique et al. (2024; see also Friston 2024; Manrique & Walker 2023), the unpredictability of innovations (i.e., appearance, disappearance, and reappearance of hand axes) arises because idiosyncratic behaviors—though sometimes leading to innovative outcomes—were often not socially reinforced within early small genus *Homo* groups. Since these behaviors were not consistently adaptive or transmitted, they frequently failed to establish lasting traditions. From this standpoint, the FEP operates on two timescales—ontogenetic (during an individual's lifetime) and phylogenetic (across generations)—and generates an itinerant movement through a space of possible behavioral states. When an individual encounters a "state" (e.g., a novel stone-flaking technique) that their internal model did not predict, they experience a cognitive surprise—a significant prediction error. This surprise could drive them to explore new behaviors, resulting in isolated technological innovations. In small, loosely connected groups with limited communication and short lifespans, novel behaviors often fell outside the group's "zone of bounded surprisal" (ZBS) (Manrique et al. 2024)—meaning they were perceived as too unpredictable or irrelevant to be retained. The ZBS is the range of surprisal within which a single agent or a group of active inference agents can operate without being overwhelmed by excessive cognitive surprise; if an event generates too much surprise, it is discarded.

Technological innovation is tightly linked to the ZBS. The innovator is the group member who pushes beyond the ZBS, thereby failing to meet the group's expectations and increasing surprise—that is, anxiety, uncertainty, sense of danger, and fear of greater energetic cost—for both the group and themselves. From a natural selection standpoint, innovation is risky—especially for small Paleolithic hominid bands, where non-orthodox behaviors that violated established behavioral priors could jeopardize survival. To outside observers, innovation is the most surprising behavior. The real challenge, however, lies in observers' ability to recognize such behavior as significant novelty and then enlarge their own ZBS by learning and imitating it (Manrique & Walker 2023, 9–10). Neurological development—which unfolds nonlinearly—is crucial for understanding both innovation itself and how it spreads through groups. Under the FEP, a key factor is the precision-weighting mechanism: an innovator's brain assigns lower precision to old priors, relaxing their influence and allowing larger prediction errors—and thus prior revision—to occur (Manrique & Walker 2023, 11–12). Manrique and Walker (2017) link this attentional mechanism—and the resulting capacity to tolerate high levels of cognitive surprise—to working memory processes that distinguish humans from chimpanzees. In practice, the width of this ZBS determines how well a group can recognize and transmit behavioral innovations, and it is directly tied to the cognitive capacities of its members. According to Walker (2025), a broader ZBS emerged during the late Miocene in small social units of mosaic-landscape–dwelling hominins than in woodland-dwelling paninans.

The interesting aspect of Manrique and Walker's (2023) analysis is that the innovation itself (the discovery) is of little relevance—it is, in fact, the failure of one member of the group to "keep uncertainty at bay" (i.e., a failure in synchronization of the generative models), and not a manifestation of greater computational power. What is truly important is the role of the observers of the innovation, that is, those who see the unorthodox behavior (i.e., the one that deviates enough from the species' phenotypical responses), become aware of it, and therefore imitate it (Manrique & Walker 2023, 17).

> The information-theoretic notion of surprisal (a.k.a. self-information) is regarded, for practical purposes, as equivalent to cognitive surprise. . . . Lesser or greater tolerance for cognitive surprise corresponds, respectively, to a narrower or broader zone of bounded surprisal (ZBS). Intolerance reflects a failure to model or infer counterfactual outcomes of surprising phenomena (e.g., novel or unintended behaviors, whether one's own or others'), due to the absence of an (evolutionarily endowed) generative model. For instance, monkeys may die from failed high-branch grasps, lacking priors for such slips. Plausibly, cognitive surprise arises from an evolved neurobiological propensity to explore

> unorthodox affordances in natural or social environments. The ZBS incorporates a temporal-mnemonic dimension: surprise contrasts with a pre-surprised baseline state, implying—even unconsciously—a capacity for that prior state to be challenged. (Walker 2025, 9)

Going back to the hand axe case, even when an individual discovered a more efficient way to shape stone tools, their peers might not have recognized its utility or lacked the generative models needed to imitate and propagate the technique. "Idiosyncratic personal attainments, ostensibly reached, as it were, atop a 'ladder' of technical abilities, could well have been disregarded by friends and family alike, who failed to imagine or articulate possible advantages that might enhance their survival" (Manrique et al. 2024, 52). Thus, the discontinuities in the archaeological record reflect a fundamental tension between exploratory (epistemic) and exploitative (pragmatic) behaviors. While individuals occasionally engaged in exploratory behavior that expanded the range of possible tool-making techniques, the lack of sustained social reinforcement meant that these innovations often disappeared, only to reemerge later under different conditions when another individual, independently, entertained similar generative models.

Manrique et al. (2024; see also Manrique & Walker 2023) highlight both the importance and complexity of leadership in technological development. They confirm that the FEP provides a powerful conceptual framework for explaining technological advancements during the Pleistocene. For an innovation to be adopted and further developed, it must remain within the group's ZBS. This idea connects to Heins et al. (2024), who examine leadership dynamics—specifically, the role of more informed agents—within a group of active inference agents interacting with one another. As mentioned above, Heins et al. (2024) argue that if only a small number of agents are informed, their influence is too weak to enable effective group coordination. Conversely, if too many agents are informed, the group becomes overly reactive to target noise, reducing efficiency and increasing the risk of fragmentation, as individuals may independently pursue their own paths toward the target. The optimal balance occurs when there are enough informed agents to guide the group while maintaining social cohesion, thereby optimizing accuracy in reaching the target by effectively regulating information flow—the FEP helps us to better define these mechanisms. It is plausible to assume that, in the Pleistocene, informed agents—those with more advanced generative models—were too few to significantly shape group dynamics. This scarcity may have contributed to the fragmentation of technological development during that period. The opposite scenario is equally problematic: an excess of leaders leads to group fragmentation, ultimately resulting in failure to achieve the goal.

Now, I believe this raises a further issue: How does competition between active inference processes occur? Assuming the presence of multiple informed agents within a group, which generative model is likely to prevail, and why? The model proposed by Friston and Frith (2015b) explains the convergence and synchronization of models, but it does not address the competition and selection of models—an aspect that is even more fundamental.

Assuming informed agents share broadly similar generative models—plausible if they compete for the same goods and resources—two interrelated factors become decisive: temporal depth and precision modulation (this reconnects the argument with the model developed by Pezzulo et al. 2021.) Temporal depth refers to the extent to which an agent can model the distal consequences of their actions. Greater depth enables the simulation of extended counterfactual trajectories, facilitating more refined policy selection and the minimization of EFE over longer time horizons. However, deeper models are computationally costly and increasingly susceptible to uncertainty propagation (this also implies that such agents develop more technology, which in turn makes the environment more salient and uncertain at the same time). Thus, competitive advantage emerges not from maximal depth per se but from the capacity to optimally regulate temporal depth in response to environmental volatility and task demands (in this case as well, the FEP provides a useful framework for formalizing the dynamics at play). Precision modulation, on the other hand, involves the dynamic weighting of prior beliefs. Agents that assign excessive precision to prior expectations risk becoming epistemically rigid, unable to accommodate unexpected outcomes or novel affordances (Manrique & Walker 2023). Conversely, agents capable of flexibly attenuating the precision of earlier priors—particularly under conditions of uncertainty—can maintain epistemic openness, enabling more adaptive model updating and improved alignment with environmental and social contingencies. These two dimensions are mutually enabling. Agents capable of attenuating the precision of earlier priors are better equipped to extend their temporal depth without overfitting past experience, enabling them to remain responsive to emerging affordances while sustaining anticipatory coherence. Such agents not only adapt more effectively to dynamic environments but also are more capable of engaging in niche construction, shaping their surroundings in ways that further reduce future uncertainty.

4.3.3 THE AREAS OF TECHNOLOGICAL INNOVATION

The ZBS on its own does not tell us how brand-new artifacts can spread in a group—it only specifies the window of "surprisal" in which agents can recognize, assimilate, or discard others' actions. Now, let us assume we have a finite number of agents whose generative models share a certain degree of similarity. A balanced number of leaders in the group thus creates the conditions for technological innovation. It is plausible to think that once this balance is achieved, the leaders will be those who select the "winning" technologies, that is, which innovation produced by non-orthodox behaviors—that is, outside the ZBS—will become a recognized innovation. Thus, an area of technological innovation (ATI)—which is based on ZBS—is created within the group where synchronization is stronger and faster, and based on richer generative models. ATI refers to the part of group leadership (i.e., their artifact-mediated generative models and the related free energy minimization) that designs and selects the most advanced artifacts for the group's purposes, thereby influencing other group members. Inevitably, peripheral agents (i.e., the learners, those outside the ATI) will begin to imitate and learn from the leaders and their technologies. This leads to two effects: (1) the raising of the cognitive threshold for the group—this means that choosing between different policies (and therefore between different artifacts) becomes increasingly difficult, that is, computationally and physically costly, and (2) an increase in leaders, that is, agents with advanced generative models according to that specific technology. Artifacts minimize free energy across the entire group—not just for the leaders—and thus also speed up learners' acquisition times. Moreover, different technologies (or clusters of technologies) might correspond to different ATIs, and each ATI has its own degree of reach or saturation (i.e., the point at which every group member becomes an informed agent, or leader, by acquiring and mastering an advanced generative model relative to that specific technology). Some technologies are saturated (e.g., hammers: while designs may vary, almost everyone knows how to use one—less surprise), whereas others are far from saturation (such as certain programming languages or spacecraft). The rate at which an ATI reaches saturation depends on two factors: the similarity among participants' initial generative models and the learners' learning speed.

We can therefore distinguish three aspects:

- *Saturable ATI:* Initially behaves like an open-ended loop: core grows, threshold rises, scope widens. Then hits diminishing returns: peripheral agents catch up; artifacts no longer off-load novel inference; group threshold flattens; ATI winds down or shifts into maintenance mode.
- *Open-ended ATI:* Never plateaus: as soon as one new generative model is mastered, a richer one emerges. Leaders and learners remain in a perpetual upward spiral, and ATI continually spawns successor ATIs without ever settling.
- *Leadership dynamics:* In saturable domains, leadership can transition into stewardship and curation once innovation winds down—fewer new leaders will emerge, and the balance tilts toward maintaining shared models. In open-ended domains (e.g., language, music), leadership remains a process of pushing boundaries—new leaders continuously emerge at each frontier, and old ones must adapt or cede ground.

Assuming (1) that there is a certain degree of similarity among the group's generative models and (2) that there is a structural coupling between the artifact's design and the generative model of the designers who conceived and built it, it follows reasonably that an ATI's saturation rate depends on the complexity of designer's generative model, and on how this influences the learners' learning speed. In active-inference terms, model complexity can be seen as the informational cost of updating a learner's internal model to account for artifact-mediated sensory inputs—that is, how much a learner must adjust their model to match the designer's (i.e., the true posterior distribution). A more complex generative model (and thus related artifacts) imposes a larger informational cost on learners, because they must make bigger or more numerous updates to their priors to align with the generative model embodied by the artifact. In a nutshell, the faster agents can pay that informational cost (learning speed), and the lower that cost (model complexity), the quicker the ATI will saturate.

The rate at which an ATI saturates is directly related to the group's ZBS—but this relation should be further clarified. The ATI is the innovation engine that arises once a ZBS is established, and a balanced leadership core exists. It applies to a smaller subgroup of people. While the ZBS defines the limits of what can be learned or adopted, the ATI is the mechanism that brings truly novel ideas—initially beyond those limits—into common use and eventually integrates them into the group's accepted repertoire. Furthermore, the speed at which learners absorb a new artifact-generative model is not constant; it typically follows an S-curve rather than a simple exponential. When an ATI first emerges, only a handful of leaders have mastered the new generative model. Peripheral agents get limited opportunities to see, use, or discuss the artifact, so their effective learning rate is low. As more agents join the core, they become additional "teachers." Each new core member propagates the original generative model (or its new versions) to their peers, so the per-agent learning rate increases. More copies, tutorials, discussion forums, best practices—and even derivative tools—appear, further lowering the cost of belief-updates. As most agents converge on the designer's generative model, there is less novel inference left to do; each remaining update fixes only minor discrepancies. Pushing learners beyond a certain proficiency can trigger diminishing returns—confusion, fragmentation into divergent submodels, or simply fatigue.

4.4 CONCLUSION

This chapter aimed to outline a roadmap for developing an FEP-based design methodology by drawing on the Beck and Ramstead (2025) and Smith et al. (2020) models. Building this model led to framing the relationship between the end user and the artifact in terms of synchronizing their generative models. At the end of the chapter, I introduced the idea of the area of technological innovation, which expands on the concept of ZBS put forward by Manrique et al. (2024). Understanding the nature and mechanisms of innovation is essential not only for design but for all of technology studies. From this perspective, the FEP once again offers a useful conceptual framework for understanding the development of human behavior in relation to the creation and use of artifacts. This also opens new research avenues: Can institutional design—that is, creating new Markov blankets around social groups—promote the development of ever more specific and effective areas of technological innovation?

Now, the claim that "design is engineering Markov blankets" may sound obvious—even trivial. It is not. In fact, we are not talking about a material, tangible, corporeal boundary but a statistical, inferential one. A Markov blanket is not just an outline; it is a frontier defined by conditional independencies. Saying that design creates Markov blankets means the designer is not merely assembling materials—they are engineering the very channels of information and inference of the user. They are actively shaping the probabilistic dependency structures that govern inferential interactions between an agent (the user) and the environment. That is a fundamental shift from traditional design:

1. It replaces subjective metrics with a unifying physical principle. Traditional design leans on usability principles and user-centered iteration and judges success with multiple—often vague—metrics like satisfaction, conversion, and time on task. The chapter's thesis proposes replacing all of that with a single, quantifiable objective: the co-minimization of free energy. The goal is no longer to make the user "happy," but to reduce costly surprises and the gap between their predictions and what they perceive. This is not a trivial claim; it is a radical proposal.
2. It makes design falsifiable and rigorously measurable. A trivial thesis does not provide new instruments of measurement; this one does. The chapter insists that if you are truly engineering a Markov blanket, you should be able to demonstrate it empirically. You can run tests to verify that the internal state is statistically independent of the external state, given the sensors and actuators. This introduces an almost scientific level of rigor and falsifiability into a discipline that often lacks it.
3. It yields a concrete methodology. A trivial thesis does not inspire new methods; this one powers the chapter's entire methodological roadmap:

   - Using the Beck and Ramstead algorithm as a starting point in order to actively forge statistical boundaries.
   - Integrating Smith et al.'s structure-learning model to explain ideation and the generation of new priors.
   - Defining three design mechanisms: precision crafting, curiosity sculpting, and prediction embedding.

5

SMARTPHONES, ROBOTS, SMART CITIES, AND HOLOGRAPHY WITH DENSE WIRELESS NETWORKS

This chapter examines several case studies. In each case, the relevant Markov blankets and generative models of the designer, user, and artifact are clearly outlined—while preserving a certain level of detail. The discussion remains nontechnical, so I will not delve into the mathematical specifics of each example. Instead, the aim is to reinforce, through conceptual analysis, the design and technology-understanding framework introduced in previous chapters and to provide a conceptual foundation for future formalizations. As noted, one benefit of reframing the human–machine relationship with active inference is that it lets us define Bayes-optimal (within-model) behavior for agents—relative to their generative models, preferences over outcomes, and chosen variational approximation—subject to environmental and resource constraints. In turn, this yields substantial advantages for design. In the appendix, I introduce the DAU software package, which models the relationship between the designer, the artifact, and the user based on active inference principles. The package is just a prototype and is freely available online. While it is initially designed around the smartphone case study, it can easily be adapted and extended to other cases. It can serve as a practical resource for designers to start getting accustomed to active inference, enabling them to simulate user–artifact interactions, mutual adaptation, and potential design strategies based on the FEP.

5.1 THE SMARTPHONE

A smartphone is a sociotechnical system involving various agents (designers and users) and technical components. It is a network (Hobbis & Hobbis 2024). I propose describing this network as a set of nested Markov blankets and relative generative models. Each active inference agent operates as an inferential machine, interacting with others through its Markov blanket. The user and the designer are not directly connected; instead, they interact exclusively through the artifact.

5.1.1 METHODOLOGICAL CAVEATS

In this context, the notion of nesting refers not to a fixed spatial hierarchy but to a temporal and inferential interdependence among agents operating at different timescales. Each agent—designer, artifact, and user—can be modeled as possessing its own Markov blanket and generative model. What makes these blankets "nested" is the way their informational boundaries and inferential dynamics interlock over time. The designer's generative model typically functions at a high level of abstraction and over longer timescales, encoding expectations and affordances that are embedded into the material and algorithmic structure of the artifact. The smartphone, in turn, instantiates these abstract priors through concrete patterns of behavior, guiding and constraining the user's generative model in moment-to-moment interaction. Crucially, this influence is not unidirectional: the user's actions—whether through gestures, decisions, or patterns of engagement—generate data that feeds back into the artifact and, eventually, into the designer's future models and decisions (see figure 5.1). In this sense, nesting is both hierarchical and dynamic: the generative models of designer, artifact, and user form a multilayered structure in which each layer predicts and is partially predicted by the others. Temporal nesting thus captures the structure of hierarchical active inference, where models at slower timescales constrain those at faster ones, and feedback propagates upward to reshape higher-level priors. The smartphone mediates this inferential ecology, functioning as a site where interacting Markov blankets overlap and evolve, not in static layers, but in continuous, temporally extended loops of mutual prediction and adaptation.

As explained in chapter 3, an artifact's type of agency is determined by its Markov blanket and the complexity of its generative model. Consider the case of a video recommendation algorithm on a platform like YouTube or TikTok. Each agent can be modeled as an inferential system with its own Markov blanket and generative model, though their levels of autonomy and inferential scope differ significantly. The designer is constantly exposed to feedback from the platform: aggregated user behavior, engagement metrics, retention rates, click-through rates, and viewing durations. Their generative model is high-level and long-term: it encodes design goals and optimization criteria (such as maximizing watch time), but it does not offer direct, real-time access to the behavior of individual users. The designer influences the system through structural and parametric changes to the algorithm. User behavior is strongly shaped by the content users are exposed to, the interface design, and subtle shifts in presentation. Their generative model is situated and guided by short-term goals—distraction, entertainment, curiosity—but it is also highly adaptive and, at times, reactive. Importantly, the user typically does not model the algorithm or the designer in a direct way; they influence both the artifact and the designer indirectly, that is, through their behavior and the data they produce.

By contrast, the recommendation algorithm—the artifact—maintains a high degree of internal coherence: it selectively receives inputs (user data, interaction history, trends) and produces outputs (ranked lists of content) in a way that is operationally stable and shielded from environmental noise. More importantly, its generative model is sophisticated and continually updated. It forms probabilistic representations of user preferences, habits, and behaviors, learning from millions of data points and interactions. The algorithm predicts, with increasing precision, the likelihood that a user will click, continue watching, stop, share, or return. It functionally inherits the objectives set by the designer, translating those into optimization routines and inferential mechanisms that guide its behavior. In this system, the algorithm emerges as the agent with the greatest capacity to influence both the user and the designer. The user is directly modeled and subtly steered by the algorithm through the dynamic arrangement of content and timing of stimuli. The designer, in turn, is indirectly shaped by the algorithm's outputs: engagement metrics, behavioral patterns, and emerging trends feed back into design decisions and future system updates. It is the algorithm that selects and evaluates the data that the designer analyses. What appears, at first glance, to be a tool becomes, in effect, an agent that mediates—and in many cases, directs—the flow of attention and behavior in the whole system.

This approach is further corroborated by existing literature on active inference. Examining human behavior on social media, Bruineberg (2023) introduces the notion of "adversarial inference," a dynamic in which users and digital platforms are locked in mutual predictions yet driven by conflicting objectives. Platforms, through algorithms that learn and anticipate user behavior, aim to maximize engagement by manipulating the environment to align with their objectives—often at odds with the user's autonomy or well-being. The artifact encodes a set of predictions about the user that exert pressure to influence and modify their behavior. According to Bruineberg (2023), these algorithms are not passive facilitators of information but active agents that exploit cognitive vulnerabilities, turning digital environments into what the author describes as instances of "hostile scaffolding." Central to the argument is the recognition that distraction in the digital age is not merely about an abundance of information. Instead, it is the continuous availability of reducible uncertainty—like the infinite scroll or the ever-updating newsfeed—that creates a compulsive loop of checking and rechecking (this can be seen, from my point of view, as a form of hyper-attunement between user and artifact, a form of addiction; more on this in chapter 6). The frictionless design of digital devices, requiring minimal motor effort to switch tasks, further exacerbates this phenomenon, subtly biasing users toward distraction or obsession. This is compounded by how digital environments leverage internal triggers—such as anxiety or boredom—to create associations that tether motivational states to platform use, reinforcing habits that serve corporate engagement goals—and this also reinforces our ideas about precision crafting and curiosity sculpting. Bruineberg (2023) advocates for a "selection for action" perspective, where attention emerges as an outcome of competing action possibilities shaped by environmental structures.

5.1.2 ANALYSIS OF THE AGENTS

In this section, I identify and analyze the Markov blankets and generative models of the active inference agents involved in this case.

The designer's generative model is used to infer the hidden states of both the user and the smartphone, based on sensory data—that is, measurable information the designer collects through interaction logs and performance analytics (e.g., crash frequency, task completion time, engagement metrics, click-through rates). The primary objective of the designer is to improve the usability and functionality of the artifact (i.e., to continuously enhance the synchronization between the artifact and the user, strengthening this coupling: the "inner synchronization"; see figure 5.1). In this case, the direction is top-down: the designer operates on timescales far longer than the user's immediate smartphone interactions, and the designer's priors are much more general and complex (i.e., including more variables about other agents and actions). Hidden states attributed to the user may include frustration, satisfaction, or behavioral tendencies, which can be inferred through indicators such as time-on-task, gaze-based attention (from eye-tracking), or habitual navigation patterns. The smartphone's hidden states, from the designer's perspective, include technical features such as touchscreen sensitivity, CPU performance,

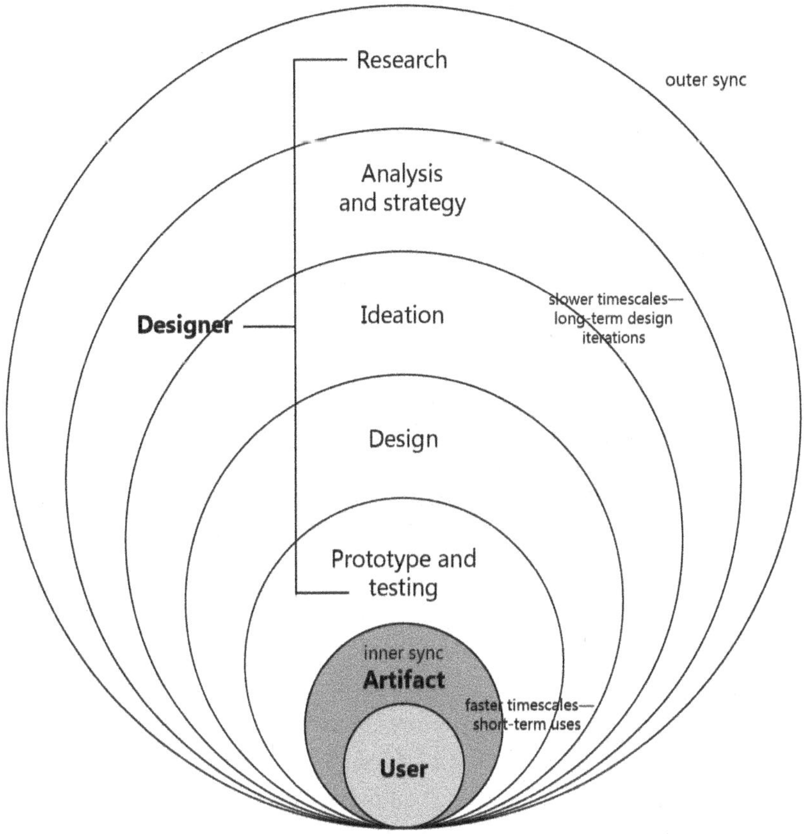

Figure 5.1
Schematic representation of a series of nested Markov blankets in the smartphone example. The framework distinguishes outer synchronization—a long-term alignment among all three agents encompassing every design phase and spanning multiple timescales—and inner synchronization, a short-term alignment between only the user and the smartphone. The designer's goal is to optimize inner synchronization based on the outer synchronization. Consequently, product iteration cycles resemble structure learning: the designer gradually refines their generative model, and, in turn, the artifact's model evolves.

software limitations, and update cycles. The designer acts on the artifact by updating or modifying its software and hardware, conducting tests, and iteratively refining it through interventions like A/B testing or interface redesign. The nature of the interventions depends on the stage of the design process.

The user's generative model, in turn, is used to infer the hidden states of both the smartphone and the designer, based on continuous multimodal interactions with the artifact—visual, auditory, and tactile. First, the user forms expectations about the smartphone's behavior (short-term predictions) and attempts (only in some cases, but it generally depends on the type of technology and the environment) to uncover the underlying intentions or logic of the designer (long-term predictions). These inferred states help the user better understand and adapt the artifact to their own goals. The smartphone's hidden states, from the user's perspective, correspond to expected system responsiveness or functional reliability, while the designer's states are interpreted as embedded general intentions, principles, or constraints guiding the artifact's design. The user acts through various means: switching apps, surfing the internet, responding to or ignoring notifications, issuing voice commands, adjusting system parameters (e.g., screen brightness, touch sensitivity), or choosing specific interface elements.

The smartphone's generative model is oriented toward inferring the user's hidden states based on their behavior during interaction. The device collects a wide range of behavioral data and, from this, generates probabilistic (short- to medium-term) predictions about the user's preferences, routines, and likely intentions. These predictions are then used to adjust the smartphone's own behavior to better align with the user's inferred model. The smartphone is capable of acting in multiple ways—for example, by modifying the graphical interface, issuing notifications, triggering vibrations, collecting data on user behavior on the internet, or dynamically adjusting usability parameters based on feedback—all with the aim of optimizing interaction and maintaining predictive alignment with the user and the designer.

Therefore, for the designer, the beliefs about hidden states—probabilistic models of user frustration, satisfaction, and device performance—are inferred solely through its observations (interaction logs, crash reports, task-completion and engagement metrics) and actions (software updates, A/B tests, firmware patches, interface redesigns), while everything outside those inputs and interventions (e.g., the user's moment-to-moment taps or environmental context) counts as external states. Likewise, the user's internal states—goals, satisfaction levels, and mental models of device behavior—are shielded by a blanket of sensory feedback (visual user interface cues, auditory alerts, and haptic vibrations) and output actions (taps, swipes, app switches, voice commands, and settings adjustments), with broader analytic pipelines and update cycles lying outside as external states. Finally, the smartphone's internal states—short- to medium-term predictions of user preferences, routines, and intentions—are conditioned only on its blanket of behavioral telemetry (touch patterns, session durations, sensor readings, network usage) and its repertoire of actions (dynamic user interface adaptations, context-aware notifications and vibrations, automatic parameter tuning, and data-collection routines), making higher-level design principles and environmental factors external to its generative model.

This informal description can be formalized in a number of ways (figure 5.2 is only an elementary representation, which uses a formalism common in the active inference literature). The key point is that formalizing the Markov blankets and the generative models help us understand the causal dependencies and feedback loops between the designer, smartphone, and user. This also enables us to answer key questions, such as: How does a designer's choice (e.g., redesigning the interface layout) propagate through the smartphone to influence user behavior? How can we describe and model this trajectory? How do user actions (e.g., frequent app switching) affect the smartphone's resource management? How do smartphone adaptations (e.g., personalized notifications) influence user engagement and satisfaction? Crucially, as said earlier, this description offers a model of sustainability: as synchronization unfolds, the three agents co-constitute a shared viability set that is not dictated from outside the systems. In other terms, sustainability means the emergent ability of designer, smartphone, and user to mutually adapt their models and actions so that their interaction remains viable over time without external control.

Precision crafting, curiosity sculpting, and predictions embedding can be seen as three interrelated mechanisms through which the designer, smartphone, and user achieve coordinated, adaptive interaction. Precision crafting involves dynamically adjusting the confidence assigned to prediction errors. The designer embeds structured constraints into the smartphone, enforcing high precision for critical feedback while allowing flexibility in more exploratory contexts. The smartphone fine-tunes precision in real time—prioritizing stable inputs (e.g., touch) and down-weighting ambiguous ones (e.g., unclear voice commands). The user adjusts precision over

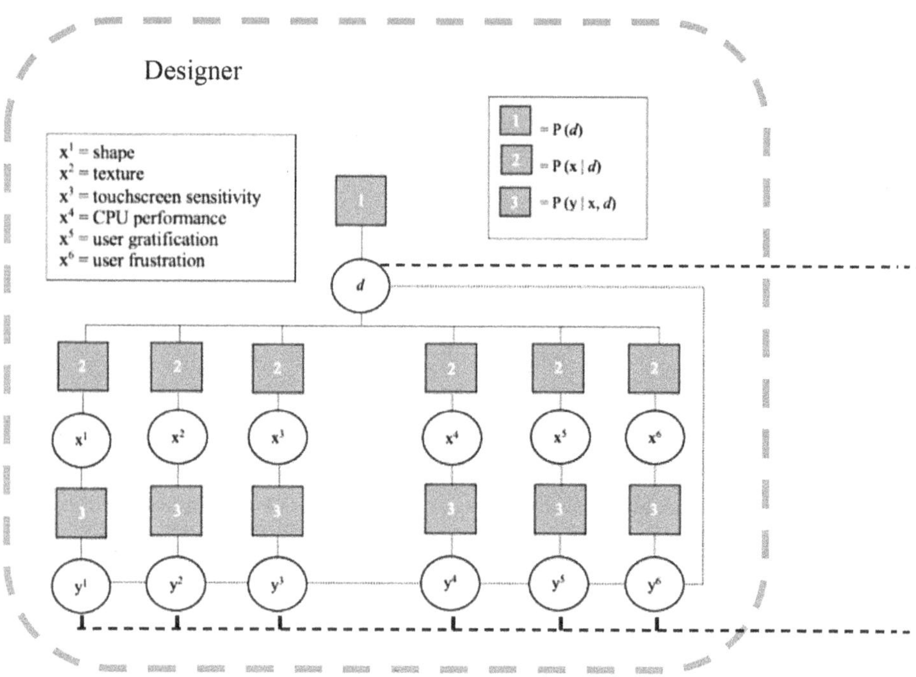

Figure 5.2
The diagram illustrates the elementary structure of the generative models associated with the designer, the smartphone, and the user. The purpose of this figure is to provide a formal intuition of the relationship among the three agents according to active inference. I use a classic model type that is widely used in the active inference literature: the partially observable Markov decision process (POMDP) (for technical details, see Nehrer 2025; Parr et al. 2022). Squares represent priors and likelihoods, while circles denote states, observations, or sequences of actions (policies). The design blanket is thicker to indicate a longer time span. The likelihood expresses the probability of an outcome (observable) given a state (hidden). If both are categorical variables, the likelihood can be formalized as a categorical distribution parametrized by a matrix. The variable *d* in the designer generative model refers to the design, which encodes a system of predictions about certain aspects of both the smartphone and the user (x^{1-6}), as well as about observations or data (y). The design also encodes predictions about policies (π^{1-3}), understood as sequences of operations performed by the artifact and implemented algorithmically. These operations may include, for instance, screen updates, vibrations, notifications, or interface changes. In this model, I adopt a temporally extended, dynamic representation, with discrete time steps t and $t+1$ (a sequences of states evolves over time). At each step, the system registers states (S, S_{t+1}) and observations (O, O_{t+1}) directly through the artifact's sensors and logging mechanisms. The artifact, therefore, maintains a generative model that predicts its own behavior in relation to the user's actions, and it continuously updates these predictions in light of new observations. The user's generative model, in turn, involves two key sets of predictions: one about the design (*d*), and one about the artifact's behavior (π^n). Based on these inferences, the user plans possible actions (*a*) to perform at each time step. Dashed lines in the diagram represent feedback relations.

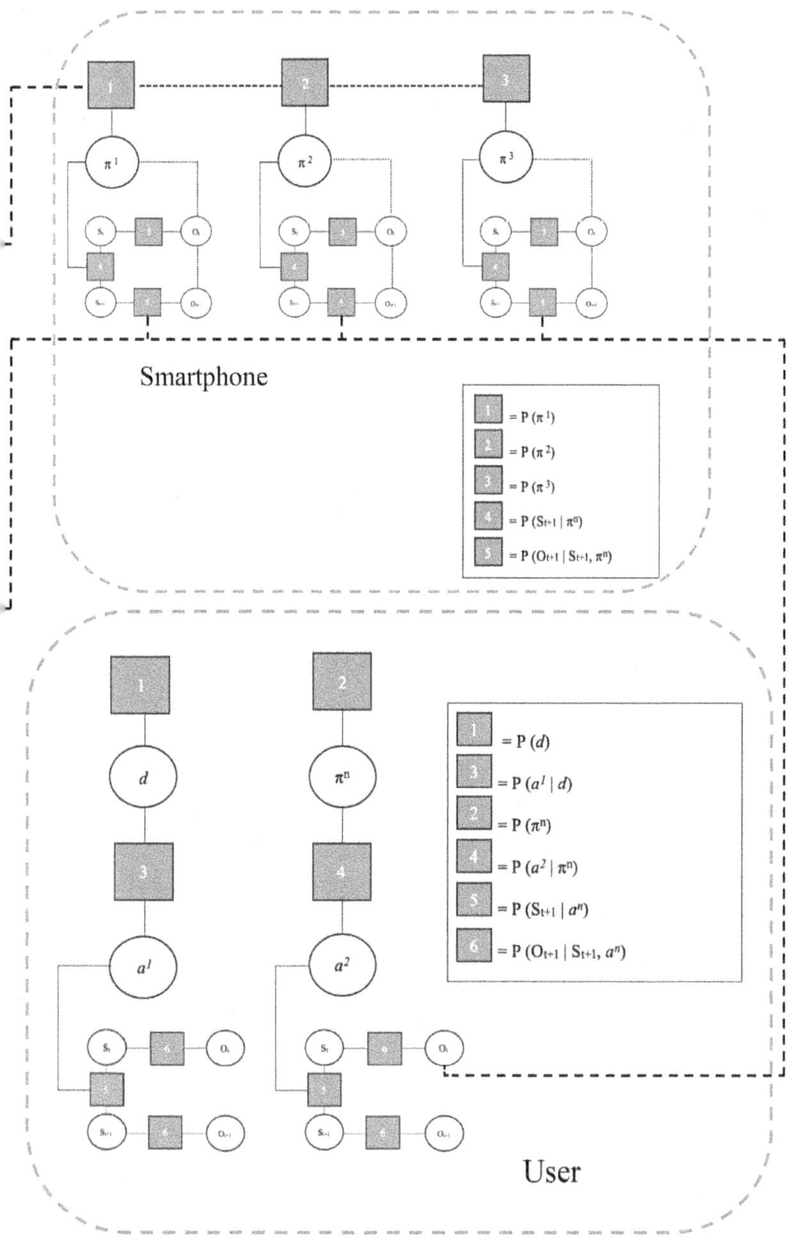
Smartphone
= P (π¹)
= P (π²)
= P (π³)
= P (St+1 | πⁿ)
= P (Ot+1 | St+1, πⁿ)
= P (d)
= P (a¹ | d)
= P (πⁿ)
= P (a² | πⁿ)
= P (St+1 | aⁿ)
= P (Ot+1 | St+1, aⁿ)
User

goals and actions, balancing focus during well-defined tasks with flexibility in open-ended situations. Curiosity sculpting fosters exploration by introducing manageable uncertainty (e.g., swiping on a dating app). Designers can embed features like adaptive tutorials or gamified elements to invite deeper engagement, while the smartphone modulates its output entropy to encourage exploration without overwhelming the user. The user responds by seeking novelty to reduce epistemic uncertainty. Predictions embedding aligns generative models: designers encode expectations into the artifact, the smartphone updates these expectations through interaction (therefore the generative model of the artifact is never completely identical to that of the designer), and the user gradually synchronizes their expectations through consistent feedback. Crucially, the designer's posteriors—the results of successive updates to their generative model throughout the design process—become the artifact's priors. Yet the artifact itself acts as an autonomous agent, capable of making predictions—of varying complexity and over different time horizons—not only about the user's behavior but also about the designer's, as the designer's subsequent generative-model updates rely on the artifact's responses.

This simple description demonstrates how active inference and the FEP provide a far more nuanced and holistic understanding of the designer–artifact–user relationship compared with traditional methods in UX design. Today, UX design is essential for boosting productivity by enhancing the usability of products. In recent years, machine learning techniques have gained significant popularity as tools to improve UX design quality (Abbas et al. 2022). However, while machine learning systems excel at automating tasks and personalizing experiences, they struggle when required to operate dynamically and adaptively. This is because they are based solely on computational principles, rather than physical ones. In contrast, active inference and the FEP offer a new framework, unlocking new possibilities for UX design. At its heart, UX design thrives on adaptability in highly uncertain settings—responding to users in real time, predicting their needs, and evolving alongside their behavior. Machine learning techniques, however, depend heavily on pre-trained models and static datasets, which limits their ability to adapt dynamically. Active inference fundamentally shifts this paradigm.

5.2 ADAPTIVE ROBOTICS

Although there are still scalability issues, current applications of active inference in robotics are proving extremely promising, especially when robots need to be designed to operate in highly complex and unpredictable environments. Da Costa et al. (2022, 1) claim that an active inference approach is an interesting framework for robotics "because it unifies state estimation, control, and world model learning as inference processes that are solved by optimizing a single objective functional: a free energy . . . as used in variational Bayesian inference." This approach endows robots "with adaptive capabilities central to real-world applications (e.g., adaptation to internal and external parameter changes)" (Da Costa et al. 2022, 1). Additionally, its strong neuroscience foundation "reduces the gap between engineering and the life sciences, thereby enhancing human-centered robotic applications" (Da Costa et al. 2022, 1). According to Da Costa et al. (2022, 6), one of the main areas of application for active inference is social robotics.

The current prevailing approach in social robotics is reinforcement learning (Maroto-Gómez et al. 2023). Reinforcement learning is a type of machine learning where an agent learns to make decisions by interacting with its environment, receiving feedback in the form of rewards or penalties, and adjusting its actions to maximize cumulative rewards over time. The agent explores different actions, learns from trial and error, and develops a strategy (or policy) that optimizes its behavior based on the observed outcomes (Han et al. 2023). Active inference differs fundamentally from reinforcement learning because it eliminates the need for rewards and value functions (Bellman optimality), which are central to reinforcement learning. This is a key logical difference: in active inference, "reward" is tied to minimizing free energy, not maximizing a generic external value function (Parr et al. 2022, 206–207; Friston et al. 2020). Reinforcement learning is based on a stimulus–response framework, where the policy consists of a set of stimulus–response associations that must be gradually learned through experience. In contrast, in active inference, the policy is embedded within the agent's generative model. As a result, in reinforcement learning, prediction error primarily concerns the reward, whereas in active inference, it concerns beliefs about states—that is, the agent predicts and updates its beliefs about the states of the world, not just the expected rewards. This logical distinction is explicitly highlighted by Friston et al. (2009), who argue that reinforcement learning is unnecessarily complex and does not guarantee adaptability. In contrast, active inference, precisely because it is not externally value-driven, allows for more dynamic and flexible adaptation.

In a nutshell, the core conceptual difference between the two approaches lies in how they understand behavior. While reinforcement learning conceives of behavior in behaviorist terms—through trial-and-error and reward—active inference treats behavior as the outcome of an inferential process. Whereas reinforcement learning controls action in a reactive manner, using stimulus–response policies, active inference controls action in a proactive way, based on a generative model of the environment (Parr et al. 2022, 210). However, the limitations of active inference should not be overlooked. Constructing accurate internal generative models can be computationally expensive, particularly for large-scale, high-dimensional tasks like real-world robotics. Reinforcement learning, on the other hand, has developed scalable methods, such as deep reinforcement learning, that efficiently manage large datasets and complex environments without requiring an explicit model of the environment. This scalability makes reinforcement learning effective in situations where building a reliable generative model is impractical or costly. Hybrid approaches increasingly demonstrate the potential for combining reinforcement learning and active inference. For instance, active inference could guide exploration by minimizing uncertainty, while reinforcement learning could optimize actions based on well-defined rewards (Tschantz et al. 2020).

5.2.1 ANALYSIS OF THE AGENTS

My goal here is not to apply active inference to robotics per se, but rather to use active inference as a conceptual framework for analyzing human–robot interaction. In this context, the artifact—the robot—exhibits a significantly higher level of complexity than a smartphone. To allow for a more grounded and precise analysis, I focus on a concrete case: the use of a robot in elder care and assistance. I assume that a social robot is "an artificial intelligent system that could socially communicate and interact with human beings" (Song & Luximon 2020).

The designer's generative model is used to predict the hidden states of both the robot and the elderly user—in essence, their behavior over the course of interaction. To do this, the designer draws on a combination of direct inputs (e.g., observation of the interaction) and indirect or mediated inputs, which may include performance metrics (e.g., task success rates), physiological indicators (e.g., heart rate variability), behavioral signals (e.g., eye-tracking data for attention or avoidance, posture, facial expression), and psychosocial assessments (e.g., Technology Acceptance Model scores). The robot itself provides additional system-level indicators such as autonomy level, turn-taking accuracy, and response latency. The designer's actions primarily involve interventions on both the software and hardware of the robot (e.g., improving sensor accuracy, reducing system latency), on the aspect (e.g., adding humanoid traits to the robot), and on the contextual environment in which the robot operates.

The user's generative model is focused on predicting the hidden states of the artifact and the designer—that is, the robot's behavioral patterns and the designer's underlying intentions. The user relies on visual, auditory, and tactile sensory input, as well as contextual knowledge. For example, if the robot is placed in a space typically used for entertainment, the user can already infer a certain function or role from this situational context. The data and predictions about the robot are the basis for a larger, more abstract inference process about the designer's intentions, that is, the overall logic of the robot. Finally, the user's possible actions can be broadly categorized into two types: collaborating or not collaborating with the robot.

The robot, due to the high complexity of the environment and the tasks, operates through two different generative models. The first is aimed at predicting the user's hidden states, which can largely be grouped into two categories: frustration (indicating failed interaction) and gratification (indicating successful interaction). To infer these states, the robot relies on a range of observable signals gathered during interaction, such as task performance, behavioral cues (eye-tracking, posture, facial expression), and vocal input. The second generative model allows the robot to infer the hidden states of both its external environment (the causes of its exteroceptive and proprioceptive perceptions) and its own internal condition as it navigates that environment (the causes of its interoceptive perceptions). This model depends on a broad set of data for spatial localization and mapping, real-time environmental perception, context understanding and interpretation, and internal system monitoring—including joint positions, energy consumption, computational load, and subsystem status.

Rather than formalizing the generative models, I am more interested in focusing on a key issue that I have only briefly touched on so far: the construction of trust and acceptance between user and robots. This is particularly important in a sensitive scenario such as elderly care. From the perspective of active inference, we can identify two complementary processes that support the development of trust in human–robot interaction. On the one hand, a user begins to trust a robot when the reliability or confidence of their predictions about the robot's behavior increases. In the user's eyes, the robot becomes more predictable—its actions align with expectations more consistently. But this is not enough. Trust also increases when the robot becomes more instrumentally valuable—that is, more useful for achieving the user's goals or satisfying their preferences. From this perspective, the robot does not simply minimize VFE through stable behavior moment by moment; it also contributes to the minimization of user's EFE by enhancing either the epistemic value (i.e., reducing uncertainty through information gain) or the pragmatic value (i.e., helping the user achieve preferred outcomes) of the user's own planning process.

Consider, for example, an elderly person who wants a cup of coffee but cannot get up—perhaps due to physical exhaustion or limited mobility. The individual knows exactly what needs to be done to satisfy this desire: go to the table, pick up the cup, and pour the coffee into it. In terms of inference, the user already possesses a well-defined policy with high confidence—there is no uncertainty about what action should be taken in that environment. What the user lacks is the physical capacity to execute that policy. Trust in the robot emerges when the robot reliably carries out the necessary action sequence to fulfill the user's preferred state—thereby minimizing EFE on the user's behalf. In this light, trust formation between user and robot can be understood as a form of reciprocal learning: a process in which the user updates priors about the reliability and utility of the robot as part of their generative model. The user progressively learns to assess and establish a certain level of confidence in the robot's behavior. The robot, on the other hand, learns to predict the user better.

Trust-building capacity is also influenced by the robot's physical appearance and the emotional responses it elicits in users. For example, as shown by Giger et al. (2025) and Song and Luximon (2020), enhancing the robot's humanoid features (e.g., responsiveness and fluid facial expressions) may contribute to increased user trust. The debate on this topic is well developed in the literature. Roesler et al. (2024) argue that there is no relationship between trust-building and robot appearance, whereas Chen et al.

(2024) suggest that a robot's personality traits significantly influence emotional trust. (See also Bhat et al. 2024; Chien et al. 2024; and Rezaei Khavas et al. 2024.)

Building on these premises, I'd now like to turn our attention to another facet that's also indirectly tied to trust. Gray and Wegner (2012) reconceptualize the so-called uncanny valley effect, shifting the explanation from visual aesthetics to cognitive expectations about mind perception. The uncanny valley describes the eerie or unsettling feeling people experience when a humanoid robot or animated character looks and moves almost like a real human—but not quite. Mori (1970) famously attributed the eerie feeling evoked by almost-human robots to a mismatch between their appearance and movement, or to subtle imperfections in their visual features—a claim that has been widely debated and challenged in the literature. Gray and Wegner (2012) reinterpret the eerie feeling in terms of theory of mind: humanoid robots are unsettling not because they closely resemble humans, but because certain features of their face or body unconsciously prompt people to attribute a mind to them—that is, to infer the presence of mental states in the robot. Drawing on the distinction between two dimensions of mind—agency (the capacity to act, plan, and exert control) and experience (the capacity to feel and think)—the authors hypothesize that uncanniness is specifically linked to violations of intuitive expectations about which entities can experience subjective mental states such as perception and feelings. Robots are typically seen as lacking experience, so when they appear capable of feeling, this mismatch generates discomfort. Three experiments supported this hypothesis. Therefore, according to Gray and Wegner (2012), the uncanny valley arises not from perceptual realism per se, but from cognitive dissonance in mind attribution. It stems "from general cognitive expectations about what should or should not have a mind, and not simply odd appearances" (Gray & Wegner 2012, 127). Nevertheless, other studies—most notably MacDorman (2024)—have challenged these conclusions. MacDorman (2024; see also Diel et al. 2025; MacDorman 2025; and Stein & MacDorman 2024) primarily critiques the methodology used by Gray and Wegner (2012), which relied solely on experiments involving written descriptions of robots rather than actual robotic interactions. In response, MacDorman (2024) introduces a new experimental setup and demonstrates that Gray and Wegner's central claim—that attributing mental states to a robot increases feelings of eeriness—is false. Instead,

Figure 5.3
Japanese roboticist Hiroshi Ishiguro with Geminoid HI-1, an android engineered to be an exact physical and behavioral replica of himself. The robot is part of the Geminoid Project, which investigates telepresence, social robotics, and the boundaries of human likeness. Geminoid HI-1 is remotely operated and replicates Ishiguro's facial expressions, voice, and micro-movements in real time. By intentionally approaching—but not fully achieving—perfect human resemblance, the project explores the psychological effects of the uncanny valley: the eerie tension that emerges when artificial agents look almost—but not quite—human. Ishiguro uses the robot to study not only human–robot interaction, but also the minimal cues required to attribute agency, emotion, or "selfhood" to a machine. *Source:* ATR Hiroshi Ishiguro Laboratories.

stimulus-driven perceptual processing, rather than the attribution of mental states, is the primary driver of uncanniness.

Now, active inference can offer a framework that integrates two previously distinct approaches to the uncanny valley: the perceptual approach, which focuses on anomalies in the stimulus, and the cognitive approach, centered on mind perception. The feeling of eeriness can be understood as the result of a failure in precision-weighting. When a person encounters a robot, they bring with them a generative model that predicts sensory inputs consistent with a nonhuman agent. However, as the robot's resemblance to a human increases beyond a certain threshold, a growing mismatch emerges between the prior belief ("this is not a human") and the sensory evidence ("this seems human"). A prediction considered reliable comes under strain. This mismatch triggers a cascade of prediction errors, both exteroceptive and interoceptive, that the system is unable to resolve. The result is epistemic uncertainty and a bodily state of ambiguous arousal—what is experienced as the uncanny or eerie feeling. In this context, the attribution of mental states (what Gray & Wegner 2012 call mind perception) does not precede the eeriness but follows from it. It emerges as a secondary, inferential attempt to resolve the ambiguity: "If it does not seem like a robot, maybe it is not just a robot—it might have feelings." From the perspective developed in the previous chapters, we can also argue that the uncanny valley—understood in a broad sense, not strictly in Mori's original formulation; consider, for example, the phenomenon of robots that are human doubles (figure 5.3)—can be interpreted as a special case of synchronization (or failed sync) between designer, artifact, and user.

5.3 SMART CITIES AND URBAN PLANNING

Why did I choose to include the smart city case in this chapter? Why is it relevant for our purpose?[1] This case was chosen to illustrate how active inference and the FEP can also be applied to large-scale, complex socio-technical systems, expanding the scope of this chapter beyond the smaller-scale interactions discussed in the smartphone and robotics cases. Unlike these systems, smart cities encompass a diverse range of agents, subsystems, and dynamic interactions, making them an ideal setting to test the scalability and adaptability of active inference on collective processes. In this case, the artifact is no longer a separate object that the designer and the user can distinguish from themselves; rather, it becomes an environment that encompasses them both. As Fox (2021) claims, active inference can provide a unifying framework for optimizing processes in large social organizations, bridging structured, repetitive systems (like industrial engineering and quality management) with more exploratory, fluid environments (such as project management). Indeed, Fox (2021) draws a compelling analogy between generative models in active inference and quality management systems used in industrial engineering.

But there is also another, deeper issue that prompted me to reflect on this case. In fact, the smart city case raises a crucial question: How can we create an environment, a space in which it is not the agent who must adapt to it, but rather the space that must adapt to the agent—and therefore must understand agent's generative models and preferences? This also connects to another equally crucial question: How do joint free-energy minimization processes influence—and in turn are influenced by—the free-energy minimization processes of individual agents within the group? What role does technology play in this relationship?

The same kind of modeling used in the previous cases can, in principle, be applied to the smart city case. However, smart cities are inherently complex, involving a vast number of interacting information and communication technology (ICT) and other technologies to improve urban living, enhance sustainability, and ensure inclusivity (Batty 2013; Halegoua 2020). A complete modeling of each agent would be too intricate. In my view, we need a different approach: treating the smart city as an autonomous Markov blanket—an emergent active inference agent.

Thestrup Waade et al. (2025) address a gap in multi-scale active inference: although many accounts propose that collectives can form a group-level Markov blanket (and thus be modeled as higher-level active inference agents), fewer studies show how to invert such a group-level model from behavioral data and relate it to the models of constituent individuals. Using a data-driven modelling approach, the authors fit a specified POMDP active-inference model to group-level observations and actions (i.e., the group's blanket dynamics) to infer group-level parameters, and then compare these estimates to the parameters governing individual agents in simulation.

Following Thestrup Waade et al. (2025), we can distinguish two types of subsystems or agents: internal agents and blanket agents (sensory and active). The sensory agents observe the environment and collect information from the individual internal agents. Those internal agents, in turn, monitor the sensory agents and act accordingly. Finally, the active agents observe the internal agents' actions and translate them into collective decisions.

How can we reconstruct, based on this approach, the generative model of the smart city? First, we identify the collective as a POMDP by specifying hidden states, observations, actions and a joint preference function. Thestrup Waade et al. (2025) lay out exactly this POMDP structure—hidden states s, observations o, actions u, and five governing matrices A–E (observation, transition, preference, initial-state, and policy priors). Second, we use variational Bayesian inference to invert that model. In Thestrup Waade et al. (2025), perception is indeed implemented as minimization of variational free energy and planning as minimization of expected free energy. Third, group-level parameters—particularly the action precision α—are estimated by fitting the POMDP to behavioral data. Finally, we validate the recovered generative model by comparing its simulated policies to the actual group decisions. Thestrup Waade et al. (2025) conduct four simulation experiments—homogeneous agents, heterogeneous α, deterministic voting, and divergent preferences—and demonstrate that the inferred group α mirrors or systematically relates to the internal agents' α under each scenario (see section 2.4). In more general terms, Thestrup Waade et al. (2025) demonstrate that any collective system—whether a board of directors, an organization, a smart city—can be understood as a single agent that builds and updates a model of the world via Bayesian inference. This implies that coordination, consensus, or conflict among members is not mysterious; rather, it predictably emerges from the free-energy minimization dynamics of the group.

Although insightful, Thestrup Waade et al.'s (2025) model has a limitation: it assumes you can clearly distinguish internal, active, and sensory agents within a group—something that is not always feasible, especially in the context of a smart city. The relationship between agents undergoes a profound shift when moving from personal technologies like smartphones to urban-scale systems such as the smart city. Although both domains involve agents with generative models interacting across nested or overlapping Markov blankets, the structure, density, and temporal dynamics of these interactions differ markedly. In the smart city, alignment between agents tends to be more fragmented and localized. Unlike discrete, tightly coupled artifacts like smartphones or robots, the smart city is a distributed, multi-agent assemblage. Its boundary is not a unified interface but a layered and heterogeneous system of data flows and control mechanisms, accessible only in part to any individual agent. Designers work with abstract, long-term models and delayed feedback, while users interact through situated, embodied experience. Each must infer the intentions and actions of the other through partial and indirect signals. Unlike the responsive, goal-directed interfaces of personal devices, the city functions as an ambient, procedural environment—its affordances shaped by spatial form, infrastructure, and regulation. Rather than interacting directly, it conditions behavior, often silently and without clear boundaries.

5.4 HOLOGRAPHY WITH DENSE WIRELESS NETWORKS

The example of radio frequency (RF)-based sensing—especially dense wireless holography—is particularly important because it shows how technology can transform our traditional understanding of active inference. Holography with dense wireless networks represents a transformative approach to wireless communication and sensing, in which large numbers of antennas or intelligent surfaces are coordinated to treat radio waves much like optical holograms, capturing and reproducing detailed electromagnetic wavefronts in three dimensions (Gong et al. 2023; Xu et al. 2024). By deploying ultra-dense arrays of small antenna elements—either as massive MIMO (Multiple-Input, Multiple-Output) panels or as reconfigurable intelligent surfaces embedded in walls, streetlamps, or devices (Shvetsov & Alsamhi 2024; Yunus et al. 2025)—networks effectively create vast continuous apertures capable of sculpting radio beams with unprecedented precision. Each tiny element can be individually tuned in phase and amplitude, allowing the network to focus energy at specific points in space, form multiple simultaneous beams, or cancel interference by creating adaptive nulls. This high degree of spatial control yields vastly improved spectral efficiency through aggressive spatial multiplexing, enabling dozens of users to share the same frequency resources without mutual interference. As mentioned, the same infrastructure serves as a high-resolution sensing platform: by measuring how transmitted waves reflect off people, objects, or environments at many different points, the network constructs real-time three-dimensional images, tracks movement with centimeter accuracy, and even detects subtle gestures or vital signs without cameras.

Such integrated sensing and communication underpin future applications from 6G mobile networks—where terabit-per-second data rates, sub-centimeter localization, and holographic telepresence become possible—to augmented and virtual reality systems that rely on ultra-low-latency links and instantaneous environmental mapping. Moreover, smart cities and autonomous transportation benefit from network-wide "radio vision," letting intelligent infrastructure assist vehicles and drones in avoiding obstacles or coordinating traffic in real time. Yet these capabilities come at a cost: building and powering thousands of tiny RF elements, handling the immense signal-processing burden of real-time channel estimation and beam computation, and synchronizing distributed surfaces to within fractions of a wavelength are formidable engineering challenges.

Now, this specific case highlights some crucial aspects for our research. The first concerns the notion of the environment. In fact, this technology creates an environment that can sense and act—continuously gathering enormous amounts of data about the agents within it, possessing more information about those agents than the agents themselves have (and this can influence their Markov blankets, given that a Markov blanket is not coextensive with the material boundaries of an object). Thanks to the small antennas, the system itself is effectively perceiving and seeing its surroundings all the time, without any separate cameras or radar. Every reflection, every shift in phase or amplitude, contributes to a live, 3D electromagnetic picture. Since this sensing capability is embedded throughout the built environment, we call it "ubiquitous perception"; that is, the system has eyes (and ears) everywhere, making the wireless network not only a communications backbone but also a pervasive, real-time sensor (and surveillant) of everything around and in it.

In a nutshell, an ultra-sensitive RF environment turns the infrastructure into a perceptual agent and forces us to think of Markov blankets as relative to the model and level of description: adding RF sensing does not violate the biological shielding of the agents, but it changes the graph and thus the relevant blankets—with implications for control, interfaces, privacy, and governance. The RF system positions itself outside the biological blanket and couples its own states to the agent's sensory and active states.

Imagine a group of agents moving in a room under this technology's surveillance (see figure 5.4); the RF sensing directly shapes the flow of information and their Markov blankets.

A Markov blanket is defined relative to a particular graphical model or factorization. If you change the variable set or add measurement nodes, what counts as the Markov blanket can change. There is not a single "true" Markov blanket independent of modeling choices or scale. More broadly, the level of description at which we define both the Markov blanket and the generative model depends on the nature of the system and on how we model the system. Even for physical systems, while the underlying dynamics may be observer-independent, our levels of description (state choices, coarse-grainings) are modeling choices. In sociotechnical systems, this dependence on the observer's generative model is even more pronounced. Crucially, this does not mean Markov blankets are not real; they are real structural features at a chosen level of description, with testable implications once that level is fixed.

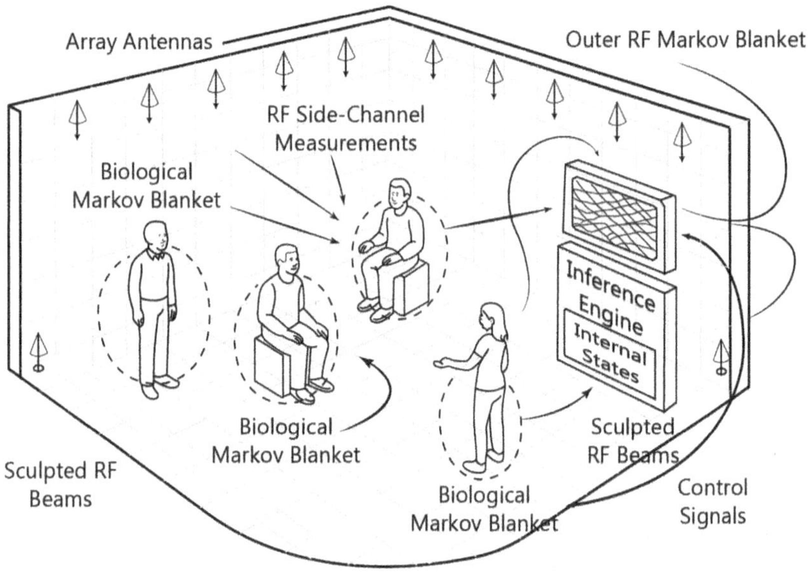

Figure 5.4
Isometric illustration of a dense wireless holographic system: array antennas form an RF Markov blanket and sculpt beams into biological Markov blankets around subjects, with side-channel measurements feeding an inference engine.

In our RF sensing environment, the "shielding" property—conditional independence of internal and external states given the blanket—of each agent's blanket does not weaken merely because sensors are added. RF sensing expands the system graph, forcing a recomputation of Markov blankets—the biology is unchanged, but the minimal separating set shifts. The augmented model simply has additional nodes and edges. After such an augmentation, one must recompute the Markov blanket in the enlarged graph; the original Markov blanket may no longer be minimal and can even fail as a separator, for example, if new colliders (Pearl 1988) or their descendants are introduced and then conditioned on. Adding RF sensors does not violate shielding; it changes what the minimal separating set is.

In this new kind of infrastructure, we are forced to distinguish different levels. For the humans inside the RF environment, RF sensors are not part of their sensory states unless those sensors are actually coupled to them. A human may still hold beliefs or priors about the RF infrastructure, but without RF transduction they receive no prediction errors from RF dynamics. By contrast, for an RF room-infrastructure agent, the RF measurements are sensory states, and in a joint model that treats the room infrastructure as an agent, its Markov blanket includes those RF sensors. If we then consider a third level—an external observer watching people move in the room—the observer's system-level blanket can subsume or overlap with the humans' blankets under many reasonable factorizations. The observer need not interact directly with people to learn about them because the infrastructure supplies richer information—often information the people themselves do not possess. RF adds data; it reduces the observer's uncertainty. It does not break the structure of the humans' Markov blankets. Designers do not add variables to a person's Markov blanket; they re-partition the environment's sensing/actuation graph. Adding sensing—especially actuation—reshapes joint blankets (e.g., humans plus infrastructure), creating new interfaces and shifting the dynamics. This, in turn, lets us specify Bayes-optimal behavior at the chosen scale and informs architecture, interfaces, control, privacy, and governance.

5.5 CONCLUSION

The aim of this chapter was to examine case studies that reinforce the arguments from earlier chapters—specifically, the idea of design as (1) engineering Markov blankets and (2) deferred, multi-level synchronization among different active-inference agents. More specifically, the case studies aimed to demonstrate (1) how active inference can explain the agency of artifacts and (2) how technology can reshape our traditional understanding of active inference. The style and methodology employed here were those of conceptual analysis; however, this work paves the way for a series of more specific, empirical studies. In my view, this can open up numerous research avenues spanning not only architectural design but also social robotics and haptic technologies.

APPENDIX: THE DAU SOFTWARE PACKAGE FOR ACTIVE INFERENCE-BASED DESIGN

The Designer-Artifact-User (DAU)[2] is a software package designed to model, simulate, and analyze the dynamic interactions between designers, artifacts, and users based on active inference. This software represents an early prototype of a design-process platform similar to Figma, Sketch, Blender, SolidWorks, Canva, or Webflow—tools designers use to craft, refine, and showcase their work across various fields. What distinguishes the DAU software package is that it is built on active inference principles. The DAU also intends to complement, rather than replace, traditional UX principles, offering a deeper understanding of how users, designers, and artifacts adapt, learn, and refine their expectations over time. While usability heuristics prioritize immediate efficiency, the DAU package focuses on short- to medium-term cognitive alignment, ensuring that artifacts evolve with the user rather than just optimizing for static interactions (see table 5.1).

The DAU package aims to explore the following research question: How do a designer's decisions influence and shape patterns of exploration versus exploitation, and information gain versus pragmatic value, in both the artifact's behavior and the user's behavior? Practically speaking, with the DAU package a designer can model their artifact as an active inference agent and then simulate its interaction with the end user as a dialogue between two active inference agents, each governed by its own generative model (see figure 5.5).

This setup allows the designer to observe how the user and the artifact co-adapt over time and to evaluate the effectiveness of each design choice. At the same time, it allows the designer to adopt an external perspective, observing how their own behavior becomes synchronized with that of the other two agents. In this way, the DAU brings together both a psychological perspective (the study of user behavior) and an engineering perspective (the study of designer behavior and the impact of their choices). From a more theoretical standpoint, the interaction and synchronization between designer, artifact, and user come before and shape affordances. The affordance can be seen as the crystallization of a synchronization between the designer, the user, and the artifact. This makes it possible to extend the viewpoint of ecological psychology: the environment is no longer just a collection of affordances (relational or human-relative properties), but, as a technological system, it becomes an active, non-human agent. This approach can be applied very effectively also in neuro-architecture (Wang et al. 2022). The building is not just a static backdrop of affordances, but an active agent that guides the human agent's active exploration. In this view, affordance emerges from the progressive synchronization among the architect, the user, and the artifact within spatial design. Starting from this basic triadic structure, it becomes possible to introduce increasing levels of

Table 5.1
Four metrics, derived from the FEP, to evaluate human–technology interaction in design. Each metric captures a key aspect of predictive alignment, uncertainty management, and adaptive engagement between users and technological artifacts.

Metric	Theoretical basis	Operational proxy	Design implication
Variational free-energy minimization	Measures the instantaneous gap between predicted sensory input (user's generative model) and actual input from the artifact	Error rates, hesitation, correction loops, eye/motor micro-adjustments	Smooth interaction, learnability, system transparency
Expected free-energy minimization	Quantifies anticipated uncertainty and reward in planned future states; reflects confidence in sustained, goal-directed use	Dropout rate, disengagement, exploratory vs. exploitative behavior	Long-term usability, trust, and commitment
Precision-weighting	Captures the confidence with which users integrate prediction errors to update beliefs; balances flexibility with stability	User reactions to novelty, adaptive behaviors, reliance on defaults vs. exploration	Support for user learning, trust calibration, context-sensitive feedback
Generative model synchronization	Assesses the inferential alignment between the designer's intended use and the user's enacted interpretation, via the artifact	Mismatch rates, unexpected use patterns, failed affordances	Intuitiveness, predictability, and interpretability of artifact behavior

complexity (e.g., more artifacts, more designers, additional layers of synchronization and precision) (see Charalambous & Djebbara 2023).

For instance, consider a smartphone design team that wants to test how different notification interfaces affect user attention and cognitive load. Using the DAU simulation package, the designer specifies three components:

1. Possible design interventions: No change (baseline), change color theme, change text size.

ENTITY INTERACTIONS

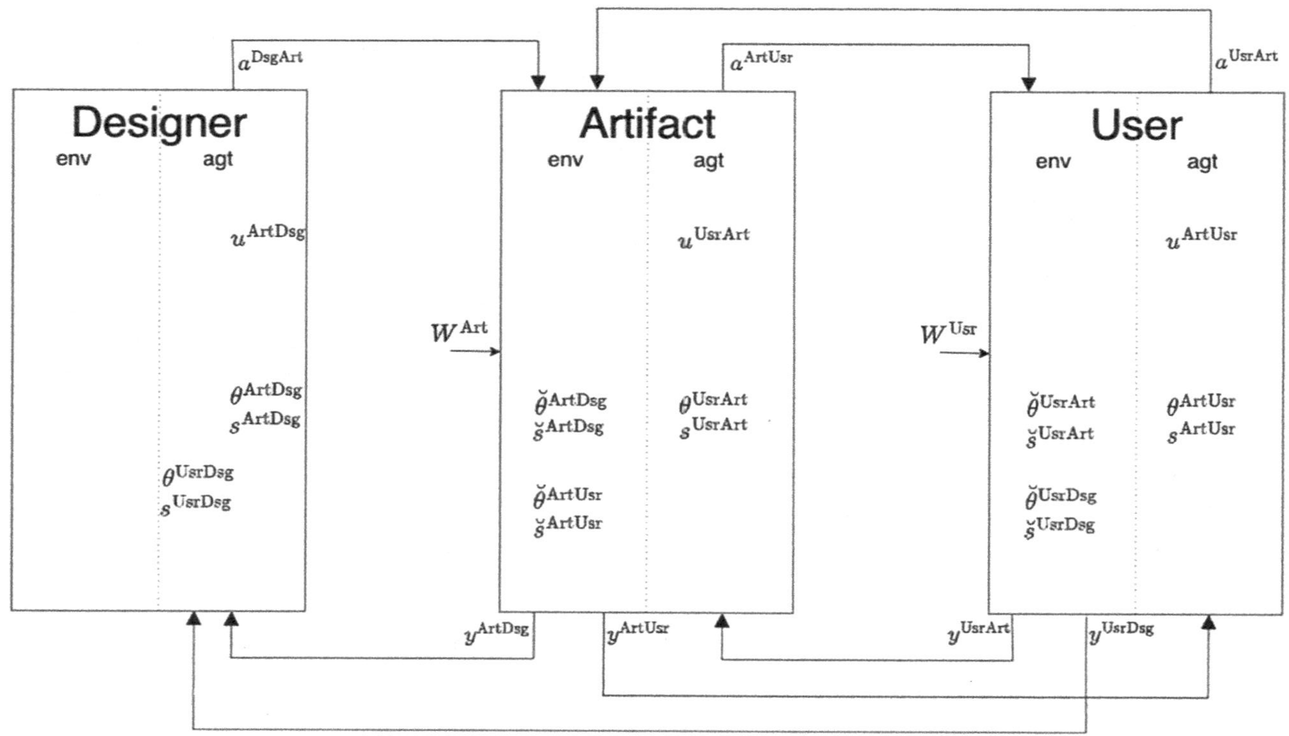

Figure 5.5
An intuitive overview of the structure of the DAU software package. The DAU frames every design problem in triadic terms, that is, as a problem of synchronization among three active inference agents: the designer, the artifact, and the user. These agents operate through different generative models and across distinct temporal and spatial scales. This framing makes it possible to derive a (within model) Bayes-optimal model of their relationship and joint behavior.

2. User: The user's cognitive states, including focused versus scattered attention (eye-tracking proxy), short, medium, or long time-on-task.
3. Artifact: The smartphone interface, which mediates between user and designer.

Once initialized, the DAU system runs a multi-step simulation of interaction. For example, if the user's attention becomes scattered after repeated notifications, the designer agent can intervene by enlarging text size or altering color theme. The simulation then updates predicted observations, such as improved eye focus or longer time-on-task.

Most multi-agent active inference systems in the literature model symmetric interactions, where multiple cognitive agents interact on equal footing. By contrast, the DAU establishes an asymmetric triad: (1) the designer, who manipulates the parameters of the artifact, guided by preferences related to the design context; (2) the user, who interacts with the artifact, shaped by their cognitive and pragmatic preferences; and (3) the artifact, which mediates the interaction and cannot be reduced to either the designer or the user. In the DAU, each agent has its own hidden states and observations, but the artifact functions as a shared Markov blanket—the layer that simultaneously separates and connects designer and user. This directly reflects concepts from postphenomenology and ANT, which emphasize that artifacts are not neutral channels but genuine cognitive and social mediators.

The full Python implementation, along with a detailed description, technical documentation, and visualization tools, is freely available as an open-source resource at https://github.com/DesignAInf/DAU-software-tool.git (DAU V2-V3 dashboard). A README file provides an overview of the project's purpose, lists its main dependencies, and offers detailed installation instructions. It also explains how to run the tool from the command line, or via the included Jupyter notebook. The document concludes with examples of result visualizations and suggestions for future development. Though this version of the prototype is based on the smartphone case study, the DAU tool can be adapted to other contexts, offering a versatile approach to modeling human–technology interactions.

6

PSYCHO-TECHNOLOGICAL ECOSYSTEMS

Technology confronts us with fundamental challenges that cannot be addressed using past experience alone. A fresh approach is necessary. This is especially clear when we consider how technologies—particularly new emerging technologies—have given rise to new social and personal practices that have profoundly transformed how we behave and interact with others, as well as with ourselves. At the same time, these new practices drive individuals to modify the very technologies they use to better meet their evolving needs (Riva & Mancini 2023). The psychology of technology focuses precisely on this human–machine loop. Understanding the relationship with technology—including the generative models underlying this relationship and the ways they are updated and evaluated—is crucial not only for grasping psychopathologies connected to technologies but also, more broadly, for comprehending how technology can shape the formation of human identity. The key idea of this chapter is that the crucial factor in mental health is the relationship with the environment—and especially with technology.

How does active inference help us understand the impact of technology on psychological well-being, particularly in relation to stress, depression, addiction, and other disorders? Generally, from the perspective of active inference and the FEP, psychopathologies such as schizophrenia, depression, anxiety, and addiction can be understood as aberrant inferences (i.e., abnormalities in prior beliefs and/or the precision-weighting; see Hesp et al. 2021; Sandved-Smith et al. 2021; and Smith et al. 2020a). This chapter expands on this idea by arguing that technology can play an active role in these processes, potentially causing or contributing to the development of these disorders. In this framework, technology-related psychopathologies are not merely the result of false inferences; rather, these false inferences emerge from distortions in the synchronization between the three key agents (designer, artifact, and user) of any technological process. From this perspective, technology-related psychopathologies are not solely the product of individual psychopathologies but arise from disruptions within a broader system of predictions.

In the first part of this chapter, drawing mainly on Feldman Barrett (2017), Panksepp (1998), and Solms (2021), I contend that interoceptive processes are hierarchically dominant in shaping cognition and behavior. My claim is that technology can actively influence interoceptive predictions and precisions. To bolster this thesis, I use Colombetti and Krueger's (2015) concept of affective scaffolds to show how technologies function as essential components of human affective niches, shaping and supporting emotional regulation. We live in *psycho-technological ecosystems*. A caveat is necessary. In this section, I am not attempting to formulate any specific thesis on consciousness. There is a growing body of literature within the active inference framework that addresses the problem of consciousness (see, e.g., Laukkonen et al. 2025; Robinson et al. 2025; Whyte et al. 2024; Wiese 2024). While the question of consciousness and the distinction between conscious states and unconscious states is undeniably intertwined with the themes explored in this chapter, my main focus here is on the psychopathologies that arise in the human–technology relationship and their connection to design. Rather than engaging in the broader philosophical debate on the nature of consciousness, my aim is to investigate how technological mediation reshapes affective dynamics, sometimes in ways that lead to dysfunction or maladaptation.

In the second part of this chapter, I further develop this approach by analyzing two case studies: depression and addiction. The main questions are: How might technology induce these forms of psychopathology, and how might it be used to treat them? What design choices can we implement to address these conditions?

In the third section of the chapter, I shift my focus to a different theme—one that may seem unrelated to the previous discussions but is, in fact, deeply connected: the concept of mentalizing. This shift allows me to revisit and further develop some of the key arguments developed in earlier chapters. In this section, I (1) introduce the concept of mentalizing, (2) interpret it through the lens of active inference, and (3) explain why mentalizing is essential for understanding technology. This section is deliberately experimental and open-ended. Instead of formal definitions, I offer an interpretation of mentalizing, examining its implications in a conceptual, exploratory way. Does mentalizing imply consciousness? Not necessarily. The ability to mentalize—understanding and attributing mental states to oneself and others—is not a fixed trait but a capacity that develops over time. It can manifest at varying degrees of complexity, rather than as an all-or-nothing phenomenon. Consciousness, rather than being a prerequisite for mentalizing, may instead emerge as a byproduct of highly refined mentalization processes. As Allen et al. (2008) argue, one can be fully capable of mentalizing others yet unable to do so in relation to oneself (in my opinion, this aligns well with certain findings in the active inference literature—see Whyte et al. 2024, 26–27).

6.1 TECHNOLOGY AND FEELINGS

6.1.1 THE DEPTH OF INTEROCEPTION

Let us recall what we have discussed so far: the brain is constantly simulating—predicting, building hierarchical systems of predictions, and calculating errors to gradually refine those systems. The fundamental point is that the brain does not do this to create an ever more accurate model of the world, but rather to anticipate the needs of the body it's connected to in order to keep it alive. This is the brain's primary mission—not to conduct science or gain objective understanding, but to manage the internal states of the body and to reach homeostasis. Consequently, the most crucial predictions—the ones that influence all others—are interoceptive predictions, those that concern the internal state of our organs. In other words, the prediction errors related to these predictions have the deepest consequences within the computational hierarchy of active inference.

Interoceptive predictions are the brain's mechanisms for managing the body's energy and maintaining homeostasis, a process Feldman Barrett (2017) describes as "body budgeting." These predictions, generated by subcortical structures such as the brainstem, hypothalamus, and basal ganglia, anticipate the body's physiological needs and prepare for them in advance, ensuring stability and smooth functioning. For example, before you stand up, your brain predicts an increase in blood flow to prevent dizziness, and before a typical mealtime, it triggers sensations of hunger in anticipation of needing energy. These processes operate mostly automatically, outside conscious awareness, forming the foundation of the body's regulatory systems. In contrast, according to Feldman Barrett (2017), the primary interoceptive cortex refines these interoceptive predictions by adding a layer of higher-order processing. This cortical activity integrates raw interoceptive data with prior experiences and contextual information, contributing to the construction of subjective experience from basic bodily signals. For instance, the flutter in your chest may be interpreted as nervousness before a high-stakes presentation, excitement before meeting a loved one, or fear during a suspenseful situation. By contextualizing bodily states, the primary interoceptive cortex bridges the gap between unconscious regulation and the construction of subjective experience. From this point of view, interoceptive active inference has a key function, which is to anticipate the resources needed and thus best adjust the body budget, that is, to never go into the "red." The brain makes a constant cascade of predictions to track and control the resources it has available (water, salt, glucose) and the possible "expenses" of energy to keep the body alive and avoid inflammation, that is, budget imbalances (Feldman Barrett 2017, 199–202). Budget imbalances mean more uncertainty.

In my view, Feldman Barrett's (2017) position is not incompatible with those of Panksepp (1998) and Solms (2021), although Feldman Barrett places greater emphasis on the cultural and socially constructed nature of emotions. By integrating these perspectives, it is possible to identify some fundamental structures of interoception.

Panksepp posits that emotions are fundamental functions of the subcortical brain, a characteristic shared by humans and animals alike. Emotions, or affects, are described as "ancient brain processes for encoding value—heuristics of the brain for making snap judgments about what enhances or detracts from survival" (Panksepp & Biven 2012, 31–32). These fundamental affective systems consist of neuroanatomical and neurochemical structures that are strikingly similar across all mammalian species (Panksepp & Biven 2012, 4).

Now, Panksepp (1998) identifies three distinct levels of brain activity:

- The primary process, encompassing the most basic emotional affects
- The secondary process, involving learning, behavioral patterns, and evolutionary habits
- The tertiary process, which pertains to higher-order cognitive functions, such as reasoning and planning.

The primary process activities are categorized into three domains: emotional affects, homeostatic affects, and sensory affects. Homeostatic affects pertain to internal biological cycles—such as hunger or the need to defecate—that maintain homeostasis. Sensory affects are reactions to external stimulations, including exteroceptive feelings of pleasure or displeasure/disgust triggered by sensory stimuli. Emotional affects, also referred to as "emotion action systems" or "intentions-in-actions," are the most ancient and complex. Panksepp identifies seven fundamental types of emotional systems: seeking, fear, rage, lust, care, panic/grief, and play. These systems, rooted in physical circuits located in the subcortical regions of the brain, drive specific reactions and behaviors (e.g., a rat fleeing from the smell of a predator and searching for a safer area to feed), which in turn foster learning. These systems are instinctual (automatic responses) and evolutionary (shaped by natural selection over time) (see also Solms 2021, 105–121). Panksepp asserts that these raw affects are the foundational basis of all brain activity. They influence and shape other cognitive processes, and Panksepp highlights the seeking system as particularly critical, describing it as the fundamental driver behind all animal behaviors. In psychoanalytic terms, it may be viewed as the primary source of psychic energy (Panksepp 2008, 165). In essence, emotional affects are ancient brain mechanisms for encoding values—heuristic processes that enable the brain to rapidly assess what enhances or reduces survival in real-world scenarios. These circuits, as detailed by Panksepp and Biven (2012, 75–76), interact with cognitive states in intricate ways but do not depend on higher-order cognitive functions. They serve as "a flexible guide for living" (Panksepp & Biven 2012, 43). Anatomically, they are fundamentally rooted in the midbrain and diencephalic regions, such as the periaqueductal gray (PAG), hypothalamus, and basal ganglia.

Following Solms (2021), we can distinguish two brain mechanisms that operate in unison within the most ancient brainstem structures: (1) Panksepp's emotional systems and (2) arousal, also localized in the PAG, which is distinct from the emotional systems but modulates their activity and priorities via a specific neural mechanism (Solms 2021, 128–132). In Solms' view, arousal is the essential condition for consciousness: "*The cortex becomes conscious only to the extent that it is aroused by the brainstem* . . . cortical consciousness *depends* upon brainstem arousal" (Solms 2021, 128). In other words, at its most fundamental level, consciousness is tied to arousal—whose primary neural hub is the reticular activating system—and, consequently, to feelings. Therefore, following Solms (2021), the reticular activating system generates arousal (i.e., postsynaptic modulation), which in turn activates and influences the cortex. The cortex then activates and influences the PAG—that is, Panksepp's primary emotional systems (Solms 2021, 136). In turn, the PAG and the "affective feelings" it produces (Solms 2021, 138) shape the entire cognitive process and guide the execution of related actions. And as Solms (2021) rightly underlines, feelings cannot be unconscious, as they are lived experiences—they "awaken" us. This also aligns with Pfaff (2005), who emphasizes the centrality of arousal as the fundamental driver of cognition, emotion, and behavior.

Let me clarify two important points. First, following Solms (2021), arousal is particularly stimulated when the emotional systems generate prediction errors, activating processes that evaluate the errors and their consequences across all levels of the brain's hierarchical organization. Second, following Solms and Friston (2018) and Solms (2021), we can define arousal as the mechanism of *interoceptive* precision-weighting—"the prediction errors (sensory inputs) that matter most to us come from *within*" (Solms 2021, 189). Arousal modulates confidence in interoceptive predictive models, adjusting the precision-weighting of prediction errors across the brain's hierarchical systems. "The adjustment of confidence levels is the essence of the work that is performed by modulatory arousal" (Solms 2021, 144). This process is fundamental to any form of free-energy minimization, serving as the mechanism by which strategies for restoring homeostasis are evaluated.

This is also supported by Hesp et al. (2021), who present a Bayesian model of emotional valence within the active inference framework. The authors argue that valence—our sense of feeling "good" or "bad"—emerges from hierarchical inference about the reliability of an agent's internal model. Specifically, valence corresponds to subjective fitness, an internal estimate of how well an agent's generative model aligns with environmental contingencies. This estimate is updated through affective charge (AC), a computational term that tracks fluctuations in expected precision—or confidence—in selecting the right actions. To validate this model, the authors simulate an affective agent in a T-maze task. The agent initially explores, using environmental cues to learn the location of a reward. As confidence in its model increases, it transitions to an exploitative strategy, selecting the optimal route without checking cues. When the reward location is reversed, the agent experiences negative valence, reflecting decreased model confidence. Over time, it relearns the new contingencies and regains positive valence, demonstrating that valence regulates adaptive learning and behavior. This model formalizes the connection between affect, metacognition, and action selection, offering a computationally grounded theory of emotion.

6.1.2 TECHNOLOGY AS AFFECTIVE SCAFFOLD

The literature on affective scaffolds is a growing field (Colombetti & Krueger 2015; Fasoli & Piredda 2023; Mossner & Walter 2025), and while an exhaustive review is beyond the scope of this section, the core thesis is that human agents manipulate their environment for affective purposes, just as environment actively influences the affective dimensions of agents. This interaction creates affective niches—organism–environment coupling enabling specific affective states. A classic example includes a song, photo, or film evoking particular emotions and memories. Similarly, Piredda (2020) discusses affective artifacts, such as cherished objects like a first bicycle, a doll, or a gift from a loved one. Spatial arrangements, colors, or material objects can convey trust, reliability, or personalization. This also means that mental health problems can be reinterpreted as "patterned dynamics" that emerge within subject–environment interactions through a process of self-organization. Niche construction is an intrinsic factor in mental health problems, and the role of technology is essential.

This gives us an opportunity to clarify a concept already introduced in chapter 3: that of affordance. Technically, an affordance in active inference is a bundle of predictions tied to the possibilities (i.e., possibilities of acting, perceiving, feeling) offered by material features of the environment. These predictions—interoceptive, exteroceptive, and proprioceptive—form a stable integration with environmental features, enabling agents to feel, act, or evolve. Building on the prior section, it is reasonable to assert that affective scaffolds and affordances are tied to our internal systems and therefore can influence and embody predictions about body budgeting and arousal levels. In other words, I predict that that song will make me feel good. At the same time, that song was designed to evoke a specific interoceptive response (positive or negative) and thus embodies a set of predictions about the internal reactions it will elicit. The material characteristics of affective scaffolds are not neutral; they actively confirm or disrupt interoceptive predictions. For example, the design of objects, spaces, or media can either align with or challenge users' expectations of comfort, trust, or reliability (Colombetti & Krueger 2015).

Technology reconfigures affectivity primarily by reshaping the affective scaffold in which an agent operates. From an active inference standpoint, this implies that technology alters interoceptive predictions and precisions, directly shaping arousal and core emotional affects. This claim is supported by a growing body of literature in the psychology of technology and media. Consider the example of a relationship ending through a WhatsApp message. The person choosing this method often does so because it feels like a simpler, less emotionally charged way to handle a tense situation. This approach avoids direct confrontation: a face-to-face exchange would require looking the other person in the eye, perceiving their emotions, and confronting an emotional weight that might influence decisions or even lead to reconsidering them. In this context, technology is used to dull the focus on interoceptive predictions—both one's own and that of others (Riva & Mancini 2023). This behavior reflects a psychological defense mechanism designed to avoid pain and guilt, two feelings often triggered by heightened emotional awareness during a breakup (Gillath et al. 2017). However, this same choice can inadvertently heighten the other person's focus on their emotional predictions. Being left via an impersonal medium like WhatsApp can provoke a range of intense and conflicting emotional reactions, such as anger, feelings of abandonment, despair, or, in some cases, cold and defensive detachment. By removing direct confrontation, the digital context not only denies a more balanced emotional closure but also amplifies feelings of rejection and isolation, which are central to attachment theories.

A very interesting case concerns the use of AI. Glickman and Sharot (2024) investigate how interactions between humans and AI systems influence human perceptual, emotional, and social judgments, revealing a profound feedback loop that intensifies (cognitive, racial, gender) biases in both humans and AI. Through a series of experiments involving 1,401 participants, the authors demonstrate that human–AI interactions can amplify initial biases, leading to significant emotional and social repercussions. The authors establish the unique dynamics of human–AI interactions, contrasting them with traditional human–human interactions. Unlike human peers, AI systems possess computational precision but are prone to amplifying subtle biases present in their training data. These biases often stem from the datasets used to train AI systems, which reflect inherent human prejudices or structural imbalances—but they may also reflect implicit decisions or biases on the part of the designer. Once AI systems internalize these biases, their interactions with humans can propagate and magnify them. The experiments revealed that human participants, when exposed to biased AI judgments, gradually adjusted their beliefs and behaviors in ways that aligned with the AI's biases, even when these deviations conflicted with their original inclinations. One illustrative experiment involved an emotion aggregation task where participants classified facial expressions as either "sad" or "happy." Initially, human judgments displayed a slight bias, but when participants interacted with an AI system trained on their biased data, this bias increased significantly over time. In contrast, when human participants interacted solely with each other under similar conditions, no such amplification of bias occurred. This same feedback mechanism can be explained by an imbalance in the precision-weighting mechanism: predictions and prediction errors coming from AI are considered more reliable and precise than those coming from humans and therefore receive more attention. This imbalance can lead humans to trust AI outputs more than judgments from their peers, even when AI systems are demonstrably biased.

6.2 TECHNOLOGY AND PSYCHOPATHOLOGY

6.2.1 A GENERAL APPROACH TO PSYCHOPATHOLOGY

Based on the existing literature, we can identify a general model of psychopathology in active inference. According to Friston (2017), psychopathology can be described as aberrant or false inference; "the key notion of false inference [is] the single most important psychopathology that underlines most psychiatric conditions—and many neurological syndromes" (Friston 2017, 640). Most scholars link psychopathological conditions to deficits in a specific parameter of active inference—precision—defined as the expected uncertainty, or the level of confidence an agent assigns to their expectations and sensory data, which in turn governs how prediction errors are weighted. Proper calibration of this parameter is crucial for active inference to work effectively (Pio-Lopez et al. 2022). Furthermore, Parr et al. (2018; see also Parr & Friston 2017, 2018; Van de Cruys and Van Dessel 2021) distinguish between *false positive inference* (hallucinations) and *false negative inference* (agnosia). False positive inference occurs when the brain perceives something that is not actually present, often due to overly strong or precise prior beliefs that override or distort incoming sensory evidence. In contrast, false negative inference refers to the failure to perceive a stimulus that is actually present. This can result from brain damage that disrupts sensory processing or attentional mechanisms.

Based on this, I propose to schematically identify four general conditions:

- Wrong priors—minimizing free energy can actually result in maladaptive behavior when it is based on incorrect priors (i.e., priors that do not match the environmental constraints).
- Excessive stress, which leads to a significant increase in prediction errors and a heightened need for updating. Chronic stress may lead to a rigid model (e.g., avoidance in post-traumatic stress disorder [PTSD]), rather than just an increased need for updating.
- Traumatic prediction error, meaning an error that disrupts fundamental priors (e.g., deeply held beliefs that are difficult to change). This can be connected to disorders like PTSD, where fundamental priors (e.g., assumptions about safety or self-worth) are violated. However, the term "traumatic" could be misleading if it is meant broadly—some destabilizing errors may not be traditionally "traumatic" but still disrupt priors.
- Imbalanced precision-weighting, referring to difficulties in assessing prediction errors and expectations, which can result in either ignoring certain errors and predictions or overemphasizing them to a nearly paralyzing extent. Precision-weighting allows the brain to recognize what is important and what is not important, that is, to choose the appropriate information in that context to construct an accurate generative model. Precision-weighting imbalances (i.e., neuromodulatory failures) can manifest in both hypersensitivity to errors (e.g., obsessive-compulsive disorder, paranoia) and hyposensitivity (e.g., schizophrenia, where sensory inputs are given insufficient weight). One of the fundamental reasons precision is so essential for explaining psychopathological conditions is its role in sensory attenuation and action (see chapter 1). Sensory attenuation is an important factor in many psychiatric disorders (Pio-Lopez et al. 2022, 5–6). From this perspective, we can also hypothesize that artifacts (e.g., symbols, writing, language more generally, and even architecture) enable or improve the regulation of sensory attenuation and movement (e.g., eye movements), and that they do so in sync with other active-inference agents.

This is a general framework that, of course, does not account for the influence of various social, economic, and cultural factors. It provides only a starting point for identifying the core concepts of psychopathology from an active inference perspective. Obviously, the development of a psychopathology is a nonlinear process that can have multiple causes of different types and various ways of manifesting.

The point, beyond formal classifications, is what practical utility—clinically and otherwise—these distinctions might have. I believe the answer lies in how these categories link psychological conditions to neural processes. The rigidity of priors—and thus the inability to update our generative models—is intrinsically tied to the rigidity of precisions, meaning we cannot revise our appraisal of priors or their associated errors. This rigidity of priors can also be understood through the concept of ZBS: each person has a specific ZBS that sets how much uncertainty they can tolerate—and how much anxiety it generates—during their affective niche-construction process. Expanding your ZBS is directly related to your ability to relax precisions; in turn, that ability may be tied to the idea of play as a transitional space (Winnicott 1971). From a Winnicottian perspective, the transitional space—characterized by play—allows the child to grapple with reality, mediating between imagination and real-world constraints and thereby preparing for frustration. In active inference terms, play is not just a structure-learning arena (Smith et al. 2020a); it is a process that transforms raw, meaningless interoceptive uncertainty into bounded, meaningful interoceptive uncertainty connected to a generative model. In psychotherapy, this transformation happens through interaction and synchronization with another agent's generative model—the therapist. The therapist creates the transitional space (the session) and offers the patient alternative generative models (i.e., exploring different narratives), thereby expanding the patient's ZBS. The patient then uses this synchronization (i.e., the synchronization with the therapist) to adjust their precisions and to generate and test alternative generative models. Synchronization in therapy happens first and foremost on the interoceptive level: it is a syncing of bodies (e.g., breathing, heart rate, vocal tone) before minds (i.e., the therapist and the patient). In a child's play, for example, synchronization occurs between the child's body and the toy (e.g., its color, texture, shape, and the sounds it produces). Like the therapist, the toy gives the child a means to regulate arousal and, in doing so, to learn to accurately assess their interoceptive states. The transitional space corresponds to "good synchronization"—an open, dynamic coupling that adaptively manages internal uncertainty—which, in turn, involves exploring generative models capable of minimizing free energy. In Winnicott's (1971) view, play is a restful interlude in which the child learns to navigate the shift from subjectivity to objectivity—moving from the exclusive intimacy of the mother–child dyad to a shared reality. In the play space, the child relinquishes the pressure to constantly update their generative models: errors become more tolerable and can even produce enjoyment. Play generates an almost uncontrolled hallucination aimed not at external action (though it obviously involves external movements) but at the internal realm—that is, the calibration of internal precisions.

6.2.2 DEPRESSION

Several studies reveal a strong association between excessive use of smartphones or social media and an increased risk of depression, anxiety, and related mental health issues (Oh & Heo 2024). Adolescents and young adults emerge as particularly vulnerable groups, especially those facing heightened stress, poor emotional regulation, or negative environmental factors such as unsupportive school climates or neglectful parenting behaviors like parental "phubbing" (ignoring someone because you are looking at your phone). Specific populations show higher rates of social media overuse, often as a coping mechanism to manage isolation, academic pressure, and emotional distress. Adolescents with risky lifestyle behaviors, including poor sleep, unhealthy diets, and reduced physical activity, are also more likely to develop depression and suicidal ideation in conjunction with smartphone dependence. Importantly, the psychological impact of smartphone overuse appears to be cyclical: it often starts as a maladaptive strategy to relieve emotional discomfort but, over time, exacerbates mental health challenges, fostering dependency and further vulnerability (Kabadayi 2024). Depression can also serve as a mediating factor in the relationship between smartphone addiction and other mental health challenges, explaining how excessive smartphone use can lead to broader psychological issues like anxiety and suicidal ideation.

Most studies in the active inference literature define depression as a malfunction in precision-weighting (e.g., Gilbert et al. 2022; Hutchinson-Wong et al. 2025). Crucial characteristics of depression (e.g., persistent negative mood, diminished agency, rumination, feelings of loss, social withdrawal, chronic fatigue—see American Psychiatric Association 2013 (DSM-5), 155; it is important to note that symptom clusters can vary significantly both among individuals at a specific point in time and within the same individual across different depressive episodes) are linked to "dysfunctional processes of prediction error minimization" and "distorted precision estimates" (Fabry 2020). In this framework, excessive precision weight is assigned to certain negative predictions over others, creating a self-reinforcing cycle of actions and perceptions and thus confirmations of those same negative predictions (i.e., predictions about the failure of other predictions). In other words, the processes of sampling and updating are oriented exclusively toward certain priors while disregarding others (Kube et al. 2020, 5). Fabry (2020), in particular, connects depression with the dark room puzzle discussed in chapter 1 of this book. A subject suffering from depression perceives change as impossible, rendering exploration and curiosity meaningless. Due to the dominance of negative predictions, the individual is unable to generate new uncertainty to subsequently minimize it. This leads to rumination and stasis (Ratcliffe 2015). According to Fuchs (2005), these aspects are also connected to "the experience of being detached from the outside world and of having lost the ability to act upon it." This detachment, in turn, generates suffering.

These analyses are important, but in my view, they miss the core of the issue. As Feldman Barrett (2017) emphasizes, depression cannot simply be regarded as a cognitive disorder, such as "negative thinking" or "high confidence in high uncertainty," where changing one's thoughts would suffice to better regulate emotions. According to Feldman Barrett (2017), depression arises from chronic metabolic misprediction, causing the brain and body to act as if under stress or illness, leading to fatigue, emotional distress, and increased risk of metabolic diseases like diabetes and heart disease. In essence, depression can be understood as a form of internal hallucination: the brain perceives something (e.g., an illness) that does not actually exist and responds accordingly. Incorrect predictions are not corrected, and errors go unacknowledged—it is a "surprising resistance to new information" (Clark 2023, 105). This means that the body continually prepares to conserve resources (e.g., limiting movement, increasing the need for sleep, activating the immune system) or to expend more resources than necessary to fight an illness that is not there. Sudden drops in energy and mood are accompanied by sudden bursts of enthusiasm and activity. This process generates not only negative thoughts, but also abnormalities of metabolic and immunological responses. Consequently, the only way to treat depression is to restore a balanced body budget by improving the brain's internal inferential processes and its precision-weighting. It is also important to note, as Feldman Barrett (2017, 212–213) explains, that anxiety stems from the same underlying cause: an imbalanced body budget but resulting from an opposite mechanism. In depression, some internal prediction errors are ignored, whereas in anxiety, they are overemphasized, leading to an excessive generation of predictions. This exacerbates the imbalance in the body budget, increasing internal uncertainty and resulting in pain. The central issue is always an instance of "aberrant precision-weighting" (Clark 2023, 106), where predictions are given excessive weight and errors are underweighted and repressed—or vice versa.

Feldman Barrett's (2017) interpretation, I believe, is also compatible with an alternative active inference view of depression proposed by Badcock at al. (2017) and Badcock and Davey (2024). Under this model, a mild to moderate depressive state can serve a valuable adaptive function, even though it can deteriorate into a pathological condition.

Following the framework outlined above, the interoceptive imbalance (i.e., chronic metabolic misprediction) at the root of depression can be traced to overly rigid priors and, consequently, to a constricted ZBS. This rigidity prevents the updating of both interoceptive generative models (e.g., accurately predicting one's internal states) and exteroceptive models (e.g., appropriately interacting with one's environment). In the case of depression associated to the use of technology, we could hypothesize an excessive synchronization between the user and the artifact—a feedback loop that intensifies over time, creating a personal echo chamber that ultimately drives self-destructive behavior. Two points deserve emphasis here. First, this over-synchronization may originate in specific design choices—such as an imbalance among the three mechanisms described in chapter 4. Second, designers can—and should—intervene to prevent this outcome and encourage healthier synchronization, that is, the transitional space as "good synchronization." This could also mean creating desynchronization spaces to reestablish clear informational boundaries between the artifact and the user. This entails having the ability to reshape the priors and the generative models of both the user and the artifact—to reconstruct the affective niche.

6.2.3 ADDICTION

Digital addiction represents a growing behavioral issue marked by compulsive and excessive engagement with digital devices, applications, and platforms, despite clear negative consequences on mental health, social relationships, and overall well-being (Fasoli 2021; Riva & Mancini 2023; Sharma et al. 2017; Veissière & Stendel 2018). Unlike casual overuse, technology addiction is characterized by a persistent inability to regulate usage, leading to significant distress and a conflict between immediate gratification and long-term satisfaction. This phenomenon is fueled by both intrinsic factors, such as the human propensity for instant rewards and overstimulation, and extrinsic factors, including persuasive design techniques deliberately implemented by technology developers to maximize user engagement. Sherer and Levounis (2022) distinguish four types of technological addictions: internet gaming disorder, social media addiction, online shopping addiction, and cybersex and online pornography addiction.

According to White and Miller (2022), addiction in active inference terms reflects a pathological recalibration of the brain's generative model by substances that concoct an abnormally steep error-reduction trajectory. Under normal conditions, dopamine signals the precision (confidence) placed on competing action policies—balancing the dual drives to minimize uncertainty and fulfill prior preferences (Friston et al. 2012a). Addictive drugs, however, directly inflate dopaminergic precision by mimicking a superlative rate of prediction error attenuation. The brain consequently overestimates the reliability of drug-seeking policies and underweights all alternative courses of action (e.g., socializing, work, hobbies). With repeated use, the system comes to expect that drug-induced "vertiginous slope" of error reduction; when it fails to materialize in ordinary contexts, unmitigated prediction errors manifest as distress—further boosting the relative precision of drug-focused policies. This creates a self-reinforcing loop: as non-drug domains accumulate unresolved errors (neglected relationships, goals, health), the agent is ever more compelled to return to the only policy still perceived as "successful"—drug use. Thus, addiction emerges from a dopaminergic hijacking of precision allocation that narrows the agent's behavioral repertoire into a rigid cycle of compulsive drug seeking and taking.

For instance, internet pornography acts as supernormal stimuli that hijack the brain's predictive machinery by offering an unprecedented flood of novel rewards. A single evening of online porn can deliver more sexual variation—multiple scenarios, hundreds of performers, ever-escalating fetishes—than our ancestors ever experienced in a lifetime. This relentless novelty creates an abnormally steep reduction in prediction error: each new image confirms that "we are doing far better than expected." Dopamine then over-allocates precision to the digital-porn–seeking policy, teaching the generative model to expect this vertiginous slope of error reduction. Real-world rewards (intimacy, social connection, hobbies) become comparatively underweighted and uncertain, so they are shunned in favor of the only policy still perceived as reliably effective.

This also aligns with Veissière and Stendel (2018), who contend that smartphones are defined by hyper-sociality rather than asociality. They situate smartphone use within a broader evolutionary framework. Their argument is that, because smartphones tap into—and amplify—a fundamentally human drive to socialize, they can foster addictive behaviors by heightening anticipation of social rewards. Smartphones emit a range of cues (e.g., notifications, videos, messages) that foreshadow these rewards, feeding anxieties about being seen (and judged) and about seeing (and judging) others. The pleasure—and thus the addictive pull—stems more from anticipating the reward than from the reward itself. In addition, these cues (e.g., the beep signaling a new message) and their associated prediction errors are processed with excessive precision.

These analyses raise an important point. Here, too, we can see a case of over-synchronization between users and artifacts. However, whereas depression involves over-synchronization between the user and the artifact based on negative experiences (i.e., unmet predictions), addiction does the opposite: it generates a cycle of hyper-reduction in free energy that nevertheless leads to a progressive rigidification of priors and poor adaptation. In this scenario, the ZBS expands dramatically through technology use, then contracts when that use is no longer possible or as intense.

6.3 AN ACTIVE INFERENCE FRAMEWORK FOR UNDERSTANDING MENTALIZING

Let us go back to the question I have already asked: What really distinguishes an artifact (e.g., a social network) from a natural object (e.g., a leaf)? The fact is that we humans recognize human intentionality in an object—that is, we are able to link that object to a mental state—we "mentalize" the object. This does not mean we literally ascribe that mental state to the object itself. Rather, we infer the presence of another human being somehow connected to it. As discussed in previous chapters, the designer (or the designers, or a long tradition of design acts and habits) creates the artifact based on their own generative model—and the artifact embodies and expresses that model, that agency. For a user to recognize and use the artifact, they must be able to reconstruct (consciously or unconsciously) that generative model through their interaction with it. Psychologically speaking, the designer attributes a set of mental states to the user just as the user attributes a set of mental states to the designer—and this reciprocal and deferred communication process happens through the mediation of the artifact. However, when I use a laptop or smartphone, I do not consciously consider the designer's mental state—I just use the device. Nevertheless, in order to recognize these objects as artifacts, I must first implicitly acknowledge their connection to a mental state. Artifacts are not just functional objects; they are interfaces of reciprocal mentalization. For example, to understand the Rosetta Stone, we first had to mentalize it—that is, attribute mental states not to the stone itself but to its creators. Recognizing that slab of rock as a human-made artifact with a specific purpose required us to infer the intentions and beliefs of the people who carved it. Clarifying and elaborating this act of mentalization took years of dedicated research and scholarship. But it was precisely this act of mentalization that enabled us to interpret and understand the stele.

In this section, starting from a preliminary analysis of the concept of mentalization, I will claim that technology-related mentalizing is far more complex than the simple synchronization between generative models we discussed earlier, even though it relies on that kind of synchronization.

6.3.1 DIMENSIONS OF MENTALIZING

Mentalizing is a fundamental human ability—the capacity to attribute mental states (such as desires, intentions, thoughts, and reasoning) to oneself and others based on observed behavior. In other words, mentalizing involves perceiving and interpreting behavior as driven by intentional mental states. Mentalizing literally involves the transformation of the nonmental into the mental, that is, the imaginative perception of behavior (one's own or that of others) as associated with mental states (Allen et al. 2008; Arabadzhiev & Paunova 2024; Luyten et al. 2020). In other terms, the social need for coordination and cooperation among agents necessitates postulating the existence of unobservable entities (mental states) to explain behavior. As Bateman and Fonagy (2016, 4) describe, mentalizing is "the ability to understand actions—both our own and those of others—in terms of thoughts, feelings, wishes, and desires; a distinctly human capability that underpins everyday interactions." This ability is rooted in social imagination and plays a crucial role in an individual's adaptation and integration into social groups, facilitating cooperation and communication. The concept of mentalizing itself is far from new (its foundations can be traced back to Wilfred Bion and French psychosomatics). It also connects to the phenomenon of primary intersubjectivity (Trevarthen 1979), which refers to the spontaneous mutual attunement of behaviors—such as gestures, vocalizations, and facial expressions—especially between infants and adults.

Any discussion of mentalizing must reference the three primary theoretical approaches to attributing mental states. The first approach, simulation theory, holds that we understand others by using our own cognitive and affective resources to simulate their mental states; this process involves an introspective mechanism whereby the observer models the target's experience as if it were their own (Goldman 2006). It is supported by findings on mirror neurons. Interaction theory or second-person approach shifts the emphasis from internal representation to embodied, reciprocal engagement. It argues that social understanding arises primarily through dynamic, real-time interactions that involve affective attunement, mutual regulation, and participatory sense-making (Reddy 2008). This view foregrounds the role of the body, timing, and context in shaping interpersonal understanding, and challenges the assumption that mentalizing is primarily inferential. Theory-theory, by contrast, posits that individuals construct and deploy an implicit "folk psychology"—a domain-specific, theory-like cognitive structure—that allows them to infer others' mental states from behavioral cues (Wellman 1990). It is a third-person, observer-based framework that treats mind-reading as an inferential, knowledge-based process, often relying on explicit reasoning or learned heuristics.

For reasons of space, I cannot elaborate on these distinctions here—I take mentalizing as a unifying concept of these different approaches. Instead, I want to focus on two essential aspects of mentalizing. The first, as underlined by Fonagy and Bateman, is that mentalizing exists on a spectrum—it is not a binary process that either occurs or does not. Instead, it involves different types and varying degrees that can be measured using specific methods (Bateman & Fonagy 2019). This means that mentalizing is characterized by uncertainty—it can be seen as a form of probabilistic reasoning: "We humans cannot know for sure what others are thinking, and as a result, there must (or perhaps should) be an element of tentativeness in our inferences about others' states of mind" (Bateman & Fonagy 2019, 63). Moreover, this ability can deteriorate or improve over time. According to researchers, many psychopathologies can be traced back to an imbalance or dysfunction in mentalizing, such as borderline personality disorder (BPD)—which, at its core, represents a distortion of social imagination (Fonagy et al. 2011). For example, individuals with BPD tend to over-mentalize others' emotions, meaning they assign excessive relevance to some external cues. This heightened sensitivity makes them particularly vulnerable in social situations. On the other hand, these individuals may also exhibit deficits in cognitive mentalizing (understanding others' thoughts) while still maintaining emotional mentalizing (perceiving others' feelings). We are not all able to mentalize to the same extent; "many of us have strengths or weaknesses in particular aspects of mentalizing, and most of us are more likely to struggle to mentalize in moments of stress or anxiety. All of us have experienced mentalizing lapses to a greater or lesser extent" (Bateman & Fonagy 2016, 5).

The spectrum of mentalizing is actually composed of four key dimensions:

- *Automatic versus controlled mentalizing:* The extent to which mentalizing happens effortlessly versus requiring deliberate, reflective thought
- *Self versus other mentalizing:* The balance between understanding one's own mental states versus interpreting those of others
- *Internal versus external mentalizing:* Whether mentalizing focuses on internal experiences (thoughts, emotions) or external cues (facial expressions, body language)
- *Cognitive versus affective mentalizing:* The distinction between reasoning about thoughts and beliefs (cognitive) versus understanding emotions and feelings (affective).

Therefore, although mentalizing can be sometimes conceptualized as a unitary concept, it is in fact "all about the balance between [these] four dimensions . . . , and about potential imbalances between the two poles of each dimension" (Bateman & Fonagy 2019, 39). There are different forms of hypomentalizing and hypermentalizing, as well as pseudomentalizing and prementalizing, each of which can affect these opposing dimensions. As the literature demonstrates, these variations can be empirically identified, quantitatively measured, and clinically analyzed (Bateman & Fonagy 2019; Fonagy et al. 2011). For instance, individuals with antisocial personality disorder (ASPD) may have a strong ability to understand others' mental states, but they often use it manipulatively, leveraging situations or people for their own benefit. This phenomenon can be interpreted in two ways: (1) as emotional hypomentalizing—a difficulty in accurately interpreting others' emotions and intentions, or (2) as hypermentalizing—some individuals with ASPD tend to overinterpret signals of threat or hostility in others, leading to impulsive, aggressive, or even violent reactions.

Only a well-balanced (both internally and externally) ability of mentalizing allows for the development of what Bateman and Fonagy (2019) call epistemic trust—the ability to perceive others as reliable sources of knowledge and to extract the most relevant information for oneself and others in relation to a given problem or situation. Epistemic trust is fundamental to salutogenesis, meaning the ability to benefit from one's environment and adapt to it, fostering resilience. A lack of epistemic trust can lead to epistemic freezing or epistemic petrification—a state of complete distrust, resulting in an inability to act or adapt (Bateman & Fonagy 2019, 71).

The second key aspect of mentalizing is its connection to attachment theory (Holmes 2014). Attachment plays a crucial role in this process, as early relationships with caregivers shape how we develop the ability to interpret others' minds. In this sense, an individual's personal history is a key factor. Numerous studies have demonstrated a strong connection between secure attachment and children's capacity for mentalization (Lundy 2003; Meins et al. 2002). These studies highlight the complex interplay between parental mentalization of the child, secure attachment, and the child's subsequent development of mentalization abilities. In contrast, traumatic or disorganized attachment leads to mindblindness—a diminished ability to mentalize and regulate emotions. A disorganized attachment style can severely impair mentalizing, leading to difficulties in understanding mental states and a tendency toward dissociation or paranoia. In these cases, the child perceives their caregiver as both a source of comfort and a source of fear, creating an internal conflict that cannot be resolved.

6.3.2 ACTIVE INFERENCE AND MENTALIZING

How can we translate the rich psychological literature on mentalization into the framework of active inference? And why would such a translation be beneficial? In fact, when comparing mentalization—as explained by Bateman and Fonagy—and active inference, a philosophical difficulty emerges, in my view. On the one hand, mentalizing seems to be the exact opposite of self-evidencing. Mentalizing, according to Bateman and Fonagy, involves suspending one's own predictions and mental models in order to fully "step into" those of the other—which, in a way, requires "other-evidencing." Yet it is also true that, if we consider the FEP—and in particular the concept of the Markov blanket—there can be no agency without co-agency. Agency never occurs in a vacuum; it is always context-dependent, that is, connected to other agencies. From this perspective, every instance of agency is intrinsically mentalizing—implicitly other-evidencing—as it imitates and partially reproduces other agencies. We form our priors, our preferences, and our generative models starting from other agents, by imitating their behavior; every instance of self-evidencing is therefore first and foremost about other agents similar to us.

This point has not always been fully recognized in the literature. For instance, Yoshida et al. (2008) present an active-inference–based account of how individuals model one another's intentions by nesting representations of an opponent's strategy within their own decision process, limiting this "recursive" reasoning to a finite depth to keep it computationally feasible. They show that by observing a sequence of choices in a cooperative hunting game, one can use Bayesian inversion to estimate how many levels of reasoning each player employs, demonstrating in both simulations and experiments that people typically reason about four to six steps ahead. Crucially, they prove that the same cooperative behavior produced by deep recursive thinking can equally emerge if agents instead maximize a utility that includes their partner's payoff, meaning apparent "mind-reading" could reflect evolved prosocial preferences rather than explicit inference. Although this framework offers valuable insights into understanding "mind-reading" from an active inference perspective, it is limited to strategic games—preplanned scenarios where agents have predefined roles. It cannot fully account for the psychological dimension of mentalization.

Now, if we want to translate mentalizing into the framework of active inference, we first need to acknowledge a crucial point. Active inference offers a theoretical advantage: it allows us to explain mentalizing not as the product of external rules and social conditions, but as an emergent dynamic arising from embodied inferential processes. This is exactly what the model proposed by Friston and Frith shows; two agents engaging in active inference synchronize their internal states through bodily, vocal, and motor signals—a form of embodied mimicry (Manrique & Walker 2023; Walker 2025). This interpersonal synchronization serves as a shared inferential baseline from which agents can infer one another's more abstract generative models. It is precisely when this baseline synchronization is disrupted—when a prediction error arises that cannot be resolved by lower-level, sensorimotor models—that the system is pushed to make a higher-order, counterfactual inference. For example, during a conversation, subtle coordination occurs in head movements, posture, and speech rhythm (Koole & Tschacher 2016; Tschacher et al. 2017). But if my interlocutor suddenly smiles in an unexpected way, that mismatch creates a prediction error that cannot be accounted for at the perceptual-motor level. In response, my inferential system begins generating hypotheses about the other's internal states: "Why did they smile? What are they trying to communicate? What are they thinking about me right now? Are they making fun of me?" In this sense, mentalizing is not a separate process, but the emergent result of attempts to resolve prediction errors that cannot be explained by forms of bodily synchronizations. It involves inference about the other's hidden states, carried out through a cascade of predictions that unfold across progressively more abstract and counterfactual levels of the brain hierarchy.

Therefore, in my view, mentalization is not only the synchronization between generative models, *but rather the response to prediction errors in bodily synchronization*. If we want to translate the psychological framework of mentalizing into active inference terms, we can distinguish four levels that match the dimensions of the spectrum (see above). Each level corresponds to a distinct type of prediction error arising from the basic, primary bodily synchronization described by the Friston and Frith model. Each error is assigned its own precision value, linking it to the spectrum dimension; in other words, each aspect of mentalizing maps onto a particular precision-weighting tied to predictions about mental-state attribution and the errors those predictions produce. Each of these levels corresponds to generative models that become progressively more abstract and span increasingly broader temporal scales. Alongside this, there is another deep interpersonal dimension: attachment theory teaches us that—interpreted through active inference—the generative models we use to understand others are passed on to us by them, or more precisely, we construct our own generative models by imitating those of others.

6.3.3 MENTALIZING THROUGH ARTIFACTS

What does all this tell us about technology? Let me highlight two points. First, as I have often reiterated, I am convinced that mentalization is an essential psychological dimension of any artifact—not only critical to its function, but even more so to its individual, social, and cultural recognizability. This psychological dimension is intrinsically linked to the artifact's agency; it in no way contradicts it. It is precisely because the artifact possesses agency that we need to consider its intentionality—that is, to mentalize it. From this standpoint, mentalizing an artifact does not mean literally attributing mental states to it but rather interrogating the design act behind it. Following what was said in the previous section, mentalization of an artifact arises the moment our immediate bodily synchronization with it falters. At that point, in order to reduce the artifact's complexity, we find ourselves needing to mentalize the designer behind it. In short, to mentalize a technological artifact is not to grant it autonomous consciousness, but to acknowledge that when interaction misaligns, it is helpful to "imagine" a mind behind it. In doing so, we simplify the problem by shifting the explanation onto the design act, and we encourage critical reflection on the intentions and responsibilities of those who build technology.

This brings us to another critical question: If humans mentalize through artifacts, can they also be mentalized by artifacts? In other words, can an artifact play an active role in mentalizing a human being? This is far from a trivial question. AI systems constantly mentalize us by attributing desires, beliefs, and thoughts based on predictive models (e.g., recommender systems, AI chatbots). Although artifacts lack a body or/and a mind, they can nonetheless "mentalize" us in a simulated way—detecting behavioral cues, inferring possible states, and offering reflections that foster epistemic trust and scaffold our own self-understanding. For example, a mood-tracking app that analyzes your voice for signs of anxiety and then prompts, "You seem stressed today—would you like to explore what's on your mind?" can feel genuinely attentive to your inner state and encourage deeper reflection. And this can have consequences for the human style of mentalizing. Perhaps the answer to this question lies in the next one: Do we mentalize artifacts directly? To what extent, and through what mechanisms? Are we capable of mentalizing them with the same level of sophistication as we do with other human beings?

6.4 CONCLUSION

I do not want to push this reinterpretation of mentalizing through the lens of active inference too far, as it risks veering into speculation without sufficient empirical grounding. Instead, I propose three key questions at the intersection of mentalizing and technology that should guide future research through empirical investigation.

- If mentalizing relies on bodily synchronization and affective attunement, can we truly mentalize a nonhuman agent—an artifact without a body, nervous system, or subjective experience? Or is mentalizing more flexible, extending beyond human interaction? If so, what enables this extension?
- If we mentalize artifacts like robots or AI assistants, are we expanding social cognition, or is this a distinct cognitive process? Does engaging with nonhuman entities blur the line between social cognition and technological mediation?
- Can technology enhance mentalizing, improving empathy and perspective-taking? Or does artificial agency create cognitive opacity, complicating mutual understanding?

CONCLUSION AND FUTURE RESEARCH LINES

This book can be summarized in three key theses.

First, design can be conceptualized as the engineering of inferential boundaries—formally describable as Markov blankets—that establish conditional independencies and stabilize coordinated behavior within sociotechnical systems. By grounding design in the control of inferential structure rather than in surface-level functionality, the book provides a principled and nonmetaphorical account of how technologies organize perception, action, and control. The point is that conceptualizing an artifact as an active-inference agent yields a Bayes-optimal account of its behavior within a specified generative model, preference structure, and resource constraints. By minimizing EFE, the agent selects actions that avoid undesirable states, reduce uncertainty, and seek information when useful, thereby remaining within a viability set defined by its outcome preferences. Unlike reinforcement learning, where viability must be externally engineered through rewards or constraints, active inference produces viability-preserving, uncertainty-aware behavior directly from its core decision rule. In this sense, active inference offers *a formal model of sustainability* as the maintenance of a system (i.e., the interactions between artifacts, designers, and users) within its viability set over time.

Second, agency is a functional property of systems that sustain organized inferential dynamics rather than a privilege of biological or mental entities. The book advances a minimal and operational notion of agency defined in terms of a system's capacity of self-evidencing. This account departs from anthropocentric and intentionalist models, allowing technical artifacts and infrastructures to be analyzed as exhibiting limited forms of agency, while explicitly preserving the conceptual distinction between agency, intentionality, and consciousness.

Third, technological influence on human cognition and emotion is best understood as a systemic modulation of precision-weighted inference within sustained human–artifact couplings. The book argues that many contemporary technologies participate directly in the regulation of attention, emotion, and action by reshaping the precision structures that govern prediction error and action selection. On this basis, the book proposes a framework for understanding how technological environments can contribute to the emergence and persistence of certain psychopathological patterns, without reducing them to simple technological causation. This is relevant in a design perspective.

One of the central points I have sought to show in this book is that the FEP functions as a principle that can be applied as a modeling approach for generic physical systems; at its core, it focuses on the time evolution of coupled random dynamical systems—which remain distinct yet interconnected with one another—and on modeling the temporal dynamics of this coupling itself. Of course, the risk of disguised metaphysics or the reification of models is ever-present; I have endeavored to minimize this tendency as much as possible.

The aim of the book was not to apply the FEP, but to redefine what the FEP does when it enters the territories of technology, design, and socio-technical networks. I believe that active inference not only provides a rigorous formal language but also introduces a set of theoretical assumptions that offer a new perspective on technology and its role in human life—what I refer to as a "third cybernetics." Of course, this is not to suggest that active inference explains everything about technology. The aim of this book was not to be apologetic. Active inference is not "the next grand revolution." Rather, it is a new language for describing the brain and the mind—and, by extension, behavior, action, and perception. Like any emerging field of research, active inference is still in its early stages. It needs, I think, to refine its methods and clarify the scope of its explanatory ambitions and its philosophical implications.

Furthermore, I do not think that understanding the connection between human cognition and technology through the lens of active inference is a form of reductionism. On the contrary, it is a way to move beyond the humanist illusion—an illusion that is, in itself, deeply reductionist. If, as Latour (1993d) argued, "we have never been modern," one reason for this is that we have never truly stopped being humanists—never ceased to believe in the illusory separation between humans and nonhumans. In reality, "man is in the machine as much as the machine is in the man" (Vial 2017, 126).

This book has aimed to explore not only *how* this hybridity occurs but also *why* it matters.

Obviously, reducing the complexity and depth of human behavior to the single imperative of minimizing uncertainty would be misguided. As I emphasized at the outset, active inference and the FEP form a highly complex mathematical framework that defines a set of quantities and their dynamic equilibria—equilibria that can be interpreted in multiple ways. These equilibria encapsulate fundamental mechanisms that govern the behavior of sentient beings.

Beyond opening this discussion, this book also aims to outline several potential research avenues:

1. Neurodivergence and technology: PP and active inference offer a unified framework for understanding neurodivergence, as discussed in Clark (2023). Integrating these insights into design practices could help develop more inclusive technologies that accommodate a wider range of cognitive profiles.
2. Technology and evolution: As demonstrated in the work of Manrique et al. (2024), active inference can also be applied to reinterpreting the history of technology—helping us understand how and why certain types of technologies emerged and evolved within human societies.
3. Rethinking AI: With a growing body of literature on the subject, active inference provides new tools for rethinking AI—grounding it in physics and biology rather than just computational efficiency. This shift has significant philosophical implications, particularly regarding our understanding of intelligence, autonomy, and learning.

ACKNOWLEDGMENTS

I would like to express my gratitude to the anonymous reviewers whose thoughtful feedback greatly enhanced the quality of this book. I am also deeply grateful to the entire MIT Press team for accepting this proposal and for their dedicated guidance throughout the development of this research. I thank Maxwell Ramstead, Karl Friston, Stephen Mann, Héctor Manrique, Michael Walker, Andrew Panshea, and Maria Raffa for their insightful comments and suggestions on the manuscript. Naturally, any errors or inaccuracies are entirely my own.

This work was supported by the Sector Plan for Science and Technology Education in the Netherlands.

For Carmel—with gratitude.

NOTES

INTRODUCTION

1. What do I mean by "design" in this context? I am primarily referring to UX design, drawing on the foundational works of Norman (2016, 2013, 2004), the essential Nelson & Stolterman (2012), and the more philosophical perspective offered by Vial (2017). These texts form the background for the discussion that follows. Nelson & Stolterman's (2012) main thesis is that design is central to human agency, enabling the creation of new realities through imagination, intention, and judgment. In an unpredictable world, design provides tools for addressing "wicked problems," characterized by ambiguity and complexity, which cannot be solved through traditional scientific or analytic methods.

2. https://www.noumenal.ai/post/wtf-is-the-fep-a-short-explainer-on-the-free-energy-principle.

3. https://www.noumenal.ai/post/wtf-is-the-fep-a-short-explainer-on-the-free-energy-principle.

4. See, for instance, https://www.verses.ai/news/verses-demonstrates-smart-building-energy-optimization?utm_medium=email&_hsenc=p2ANqtz--dxg5PGcBNcr3KSWLf8lY1QvJxZ7jXuT-_Jd0Rmy0U6090jQnhI6vHt-v855xu370onFh50Y4hO_nPRQZCPS2Ly7GwEpg&_hsmi=352424850&utm_content=352424850&utm_source=hs_email.

5. The distinction between first and second cybernetics lies primarily in their perspectives on control, observation, and the nature of systems. First cybernetics, developed in the 1940s and 1950s by pioneers like Norbert Wiener, Warren McCulloch, and William Ross Ashby, treats systems as externally observed entities governed by negative feedback loops that maintain equilibrium and stability. In this view, the observer remains separate, and the system's goal is to achieve predictable control, as exemplified by thermostats or biological homeostasis. By contrast, second cybernetics, emerging in the 1960s and 1970s with contributions from thinkers such as Heinz von Foerster, Humberto Maturana, and Francisco Varela, integrates the observer into the system, emphasizing self-organization, autopoiesis, and the recursive nature of feedback. Here, systems are open, adaptive, and evolve through the participatory construction of reality, acknowledging that the observer's involvement fundamentally shapes the system itself.

CHAPTER 1

1. The classic objection to this thesis is that if perception is a controlled hallucination, then anything goes. It is interesting how Seth (2021) responds to this objection by referring to Locke and the classical distinction between primary and secondary qualities. "Describing perception this way doesn't mean that anything goes, it means that the way in which things in the world appear in perceptual experience is a construction of the brain" (Seth 2021, 93–94). It is worth clarifying that prediction in PP "is not necessarily about the future. It simply means going beyond the data by using a model. In statistics, the essence of prediction is in catering for the absence of sufficient data. Whether this is because predictions are about the future—one can think of the future as insufficient data—or about some current but incompletely unknown state of affairs, doesn't matter" (Seth 2021, 108).

CHAPTER 2

1. See https://www.noumenal.ai/post/wtf-is-the-fep-a-short-explainer-on-the-free-energy-principle.

2. See https://www.ashby.info/aphorisms.html.

CHAPTER 3

1. The term "actor" is generally more widely used than "agent" in the ANT literature. However, in order not to make the terminology I use in this book too complex, I prefer to keep the term "agent."

CHAPTER 5

1. As regards the application of active inference to smart cities, see https://medium.com/@deniseholt1/building-the-future-how-active-inference-ai-and-the-spatial-web-will-transform-abu-dhabi-into-the-97185194cec1.

2. I thank Kobus and Stefan Esterhuysen for their assistance in developing the technical aspects of this tool.

REFERENCES

Abbas, A. M., Ghauth, K. I., Ting, C. Y. 2022. "User Experience Design Using Machine Learning: A Systematic Review." *IEEE Xplore* 10: 51501–51514.

Achterhuis, H. (2001). *American Philosophy and Technology: The Empirical Turn.* Bloomington: Indiana University Press.

Akrich, M., Latour, B. 1992. "A Summary of a Convenient Vocabulary for the Semiotics of Human and Nonhuman Assemblies." In *Shaping Technology/Building Society*, ed. W. E. Bijker and J. Law, 170–196. Cambridge, MA: MIT Press.

Allen, J. G., Fonagy, P., Bateman, A. W. 2008. *Mentalizing in Clinical Practice.* New York: American Psychiatric Pub.

American Psychiatric Association. 2013. *The Diagnostic and Statistical Manual of Mental Disorders, Fifth Edition, Text Revision* (DSM-5-TR).

Andrews, M. 2021. "The Math Is Not the Territory: Navigating the Free Energy Principle." *Biology & Philosophy* 36(3): 30–45.

Anscombe, G. E. M. 1957. *Intention.* Cambridge, MA: Harvard University Press.

Arabadzhiev, Z., Paunova, R. 2024. "Complexity of Mentalization." *Frontiers in Psychology*. 15: 1353804.

Aristotle. 1984. *The Complete Works of Aristotle* (ed. J. Barnes). Princeton, NJ: Princeton University Press.

Ashby, W. R. 1952. *Design for a Brain.* New York: Wiley & Sons.

Ashby, W. R. 1956. *An Introduction to Cybernetics.* New York: Chapman and Hall.

Attias, H. 2003. "Planning by Probabilistic Inference." *Proceedings of the 9th International Workshop on Artificial Intelligence and Statistics* 4: 9–16.

Badcock, P., Davey, C. G. 2024. "Active Inference in Psychology and Psychiatry: Progress to Date?" *Entropy* 26(10): 833.

Badcock, P., Davey, C. G., Whittle, S., Allen, N. B., Friston, K. J. 2017. "The Depressed Brain: An Evolutionary Systems Theory." *Trends in Cognitive Sciences* 21(3): 182–194.

Badcock, P., Ramstead, M., Sheikhbahaee, Z., Constant, A. (eds.). 2022. *Applying the Free-Energy Principle to Complex Adaptive Systems.* Basel: MDPI.

Baltieri, M., Buckley, C., Bruineberg, J. 2020. "Predictions in the Eye of the Beholder: An Active Inference Account of Watt Governors." In *Proceedings of the 2020 Conference on Artificial Life*, ed. J. Bongard, J. Lovato, L. Hebert-Dufrésne, R. Dasari, L.Soros, 121–129. Cambridge, MA: MIT Press.

Barad, K. 2007. *Meeting the Universe Halfway*. Durham, NC: Duke University Press.

Bastos, A. M., Usrey, W. M., Adams, R. A., Mangun, G. R., Fries, P., Friston, K. 2012. "Canonical Microcircuits for Predictive Coding." *Neuron* 76(4): 695–711.

Bateman, A., Fonagy, P. 2016. *Mentalization-Based Treatment for Personality Disorders: A Practical Guide*. Oxford: Oxford University Press.

Bateman, A. W., Fonagy, P. (eds.). 2019. *Handbook of Mentalizing in Mental Health Practice*. New York: American Psychiatric Pub.

Batty, M. 2013. "Big Data, Smart Cities and City Planning." *Dialogues in Human Geography* 3(3): 274–279.

Beck, J., Ramstead, M. 2025. "Dynamic Markov Blanket Detection for Macroscopic Physics Discovery." *arXiv preprint arXiv:2502.21217*.

Bell, M. A., Deater-Deckard, K. 2007. "Biological Systems and the Development of Self-Regulation: Integrating Behavior, Genetics, and Psychophysiology." *Journal of Developmental and Behavioral Pediatrics* 28(5): 409–420.

Beni, M. D. 2021. "A Critical Analysis of Markovian Monism." *Synthese* 199: 6407–6427.

Bennett, J. 2010. *Vibrant Matter*. Durham, NC: Duke University Press.

Bhat, S., Lyons, J. B., Shi, C., Yang, X. 2024. "Evaluating the Impact of Personalized Value Alignment in Human–Robot Interaction: Insights into Trust and Team Performance Outcomes." In *Proceedings of the 2024 ACM/IEEE International Conference on Human–Robot Interaction*, ed. D. Grollman, E. Broadbent, 32–41. New York: Association for Computing Machinery.

Bijker, E., Hughes, T., Pinch, T. (eds.). 1993. *The Social Construction of Technological Systems*. Cambridge, MA: MIT Press.

Bishop, C. 2012. *Pattern Recognition and Machine Learning*. Berlin: Springer.

Bishop, M., Bishop, H. 2025. *Deep Learning: Foundations and Concepts*. Berlin: Springer.

Blok, A., Farías, I., Roberts, C. 2020. *The Routledge Companion to Actor-Network Theory*. New York: Taylor & Francis.

Bogacz, R. 2017. "A Tutorial on the Free Energy: Framework for Modelling Perception and Learning." *Journal of Mathematical Psychology* 76: 198–211.

Botin, L., Berling Hyams, I. (eds.). 2021. *Postphenomenology and Architecture: Human–Technology Relations in Built Environment*. London: Lexington Books.

Bouizegarene, N., Ramstead, M., Constant, A., Friston, K., Kirmayer, L. 2024. "Narrative as Active Inference: An Integrative Account of Cognitive and Social Functions in Adaptation." *Frontiers in Psychology* 15: 1345480.

Boyce, P. 2021. "Light, Lighting and Human Health." *Lighting Research & Technology* 54(2): 101–144.

Brey, P. 2005. "Artifacts as Social Agents." In *Inside the Politics of Technology: Agency and Normativity in the Production of Technology and Society*, ed. H. Harbers, 128–145. Amsterdam: Amsterdam University Press.

Brey, P. 2006. "Social and Ethical Dimensions of Computer-Mediated Education." *Journal of Information, Communication and Ethics in Society* 4(2): 91–101.

Bruineberg, J. 2023. "Adversarial Inference: Predictive Minds in the Attention Economy." *Neuroscience of Consciousness*. https://doi.org/10.1093/nc/niad019.

Bruineberg, J., Dołęga, K., Dewhurst, J., Baltieri, M. 2022. "The Emperor's New Markov Blankets." *Behavioral and Brain Sciences*. 45: e183. https://doi.org/10.1017/S0140525X21002351.

Bruineberg, J., Kiverstein, J., Rietveld, E. 2016. "The Anticipating Brain Is Not a Scientist." *Synthese* 195: 2417–2444.

Bruineberg, J., Rietveld, E., Parr, T., van Maanen, L., Friston, K. 2018. "Free-Energy Minimization in Joint Agent–Environment Systems: A Niche Construction Perspective." *Journal of Theoretical Biology* 455: 161–178.

Callon, M. 1986. "Éléments pour une sociologie de la traduction." *L'Année sociologique* 36: 169–208.

Callon, M. (ed.). 1989. *La Science et ses reseaux*. Paris: La Découverte.

Camazine, S., Deneubourg, J.-L., Franks, N. R., Sneyd, J., Theraulaz, G., Bonabeau, E. 2001. *Self-Organization in Biological Systems*. Princeton, NJ: Princeton University Press.

Carhart-Harris, R., Chandaria, S., Erritzoe, D., Gazzaley, A., Girn, M., Kettner, H., Mediano, P., Nutt, D., Rosas, F., Roseman, L., Timmermann, C., Weiss, B., Zeifman, R., Friston, K. 2023. "Canalization and Plasticity in Psychopathology." *Neuropharmacology*. 226: 109398. https://doi.org/10.1016/j.neuropharm.2022.109398.

Carlsmith, J. 2022. "Is Power-Seeking AI an Existential Risk?" *arXiv:2206.13353*.

Castilla, N., Higuera-Trujillo, J., Llinares, C. 2023. "The Effects of Illuminance on Students' Memory: A Neuroarchitecture Study." *Building and Environment* 228: 53–78.

Charalambous, E., Djebbara, Z. 2023. "On Natural Attunement: Shared Rhythms between the Brain and the Environment." *Neuroscience & Biobehavioral Reviews* 155: 105438. https://doi.org/10.1016/j.neubiorev.2023.105438.

Chen, N., Liu, X., Hu, X. 2024. "Effects of Robots' Character and Information Disclosure on Human–Robot Trust and the Mediating Role of Social Presence." *International Journal of Social Robotics* 16: 811–825.

Cheng, T., Sato, R., Hohwy, J. (eds.). 2024. *Expected Experiences: The Predictive Mind in an Uncertain World.* New York: Routledge.

Chien, S., Lin, Y., Chang, B. 2024. "The Effects of Intimacy and Proactivity on Trust in Human–Humanoid Robot Interaction." *Information Systems Frontiers* 26: 75–90. https://doi.org/10.1007/s10796-022-10324-y

Chirimuuta, M. 2024. *The Brain Abstracted.* Cambridge, MA: MIT Press.

Chivers, T. 2025. *Everything Is Predictable.* New York: Weidenfeld & Nicolson.

Ciaunica, A., Levin, M., Rosas, F. E., Friston, K. 2023. "Nested Selves: Self-Organization and Shared Markov Blankets in Prenatal Development in Humans." *Topics in Cognitive Science.* https://doi.org/10.1111/tops.12717.

Clark, A. 2008. *Supersizing the Mind.* Oxford: Oxford University Press.

Clark, A. 2013. "Whatever Next? Predictive Brains, Situated Agents, and the Future of Cognitive Science." *Behavioral and Brain Sciences* 36(3): 181–204.

Clark, A. 2014. *Mindware: An Introduction to the Philosophy of Cognitive Science.* Oxford: Oxford University Press.

Clark, A. 2016. *Surfing Uncertainty.* Oxford: Oxford University Press.

Clark, A. 2017. "Predictions, Precision, and Agentive Attention." *Consciousness and Cognition* 56: 115–119.

Clark, A. 2022. "Extending the Predictive Mind." *Australasian Journal of Philosophy* 102(1): 119–130.

Clark, A. 2023. *The Experience Machine.* New York: Allen Lane.

Clark, A., Chalmers, D. 1998. "The Extended Mind." *Analysis* 58: 10–23.

Clark, A., Pritchard, D., Vaesen, K. 2012. "Extended Cognition and Epistemology." *Philosophical Explorations* 15(2): 87–90.

Collins, H. M., Yearley, S. 1992. "Epistemological Chicken." In *Science as Practice and Culture*, ed. A. Pickering, 369–389. Chicago: University of Chicago Press.

Colombetti, G., Krueger, J. 2015. "Scaffoldings of the Affective Mind." *Philosophical Psychology* 28(8): 1157–1176.

Colombo, M. 2019. "Learning and Reasoning." In *The Routledge Handbook of the Computational Mind*, ed. M. Sprevak, M. Colombo, 381–396. New York: Routledge.

Colombo, M. 2020. "Resource-Rationality as a Normative Standard of Human Rationality." *Behavioral and Brain Sciences* 43: e5. https://doi.org/10.1017/S0140525X19001596.

Colombo, M., Elkin, L., Hartmann, S. 2021. "Being Realist about Bayes, and the Predictive Processing Theory of Mind." *British Journal for Philosophy of Science* 72: 185–220.

Colombo, M., Palacios, P. 2021. "Non-Equilibrium Thermodynamics and the Free Energy Principle in Biology." *Biology & Philosophy* 36(41): 1–26.

Colombo, M., Wright, C. 2018. "First Principles in the Life Sciences: The Free-Energy Principle, Organicism, and Mechanism." *Synthese* 198: 3463–3488.

Conant, R., Ashby, W. 1970. "Every Good Regulator of a System Must Be a Model of That System." *International Journal of Systems Science* 1(2): 89–97.

Constant, A. 2020. "The Free Energy Principle: It's Not about What It Takes, It's about What Took You There." *Biology & Philosophy* 36: 10. https://doi.org/10.1007/s10539-021-09787-1.

Constant, A., Clark, A., Friston, K. 2019a. "Representation Wars: Enacting an Armistice through Active Inference." *Frontiers in Psychology* 11. doi.org/10.3389/fpsyg.2020.598733.

Constant, A., Friston, K., Clark, A. 2024. "Cultivating Creativity: Predictive Brains and the Enlightened Room Problem." *Philosophical Transactions of the Royal Society* 379(1895): 20220415.

Constant, A., Ramstead, M., Veissière, S., Campbell, J., Friston, K. 2018. "A Variational Approach to Niche Construction." *Journal of the Royal Society Interface* 15(141): 20170685.

Constant, A., Ramstead, M., Veissière S., Friston, K. 2019b. "Regimes of Expectations: An Active Inference Model of Social Conformity and Human Decision Making." *Frontiers in Psychology* 10: 679. https://doi.org/10.3389/fpsyg.2019.00679.

Constant, A., Tschantz, A., Millidge, B., Criado-Boado, F., Martinez, L., Müeller, J., Clark, A. 2021. "The Acquisition of Culturally Patterned Attention Styles under Active Inference." *Frontiers in Neurorobotics* (15): 729665.

Corcoran, A. W., Pezzulo, G., Hohwy, J. 2020. "From Allostatic Agents to Counterfactual Cognizers: Active Inference, Biological Regulation, and the Origins of Cognition." *Biology and Philosophy* 35(3): 32–76.

Criado-Boado, F. 2014: "Archaeologies of Space." In *Paradigm Found: Archaeological Theory—Present, Past and Future*, ed. K. Krisitansen, L. Smejda, J. Turek, 61–83. Oxford: Oxbow Books.

Criado-Boado, F., Alonso-Pablos, D., Blanco, M. J., Porto, Y., Rodríguez-Paz, A., Cabrejas, E., et al. 2019. "Coevolution of Visual Behaviour, the Material World and Social Complexity, Depicted by the Eye-Tracking of Archaeological Objects in Humans." *Scientific Reports* 9: 3985. https://doi.org/10.1038/s41598-019-39661-w.

Cronon, W. 1991. *Nature's Metropolis: Chicago and the Great West*. New York: Norton Press.

Cummins, R. 2014. "Homeostasis." In *Encyclopedia of Quality of Life and Well-Being Research*, ed. F. Maggino, 3165–3168. Cham: Springer International Publishing.

Da Costa, L., Lanillos, P., Sajid, N., Friston, K., Khan, S. 2022. "How Active Inference Could Help Revolutionize Robotics." *Entropy* 24(3): 361–374.

Davidson, D. 1963. "Action, Reasons, and Causes." *Journal of Philosophy* 60(23): 685–700.

Dayan, P., Abbott, L. 2001. *Theoretical Neuroscience: Computational and Mathematical Modeling of Neural Systems*. Cambridge, MA: MIT Press.

DeLanda, M. 2016. *Assemblage Theory*. Edinburgh: Edinburgh University Press.

Dennett, D. C. 1987. *The Intentional Stance*. Cambridge, MA: MIT Press.

Dennett, D. C. 2003. *Freedom Evolves*. New York: Viking.

Diel, A., Lalgi, T., Teufel, M., Bäuerle, A., MacDorman, K. 2025. "Eerie Edibles: Realism and Food Neophobia Predict an Uncanny Valley in AI-Generated Food Images." *Appetite* 208: 107926.

Dimakou, A., Pezzulo, G., Zangrossi, A., Corbetta, M. 2025. "The Predictive Nature of Spontaneous Brain Activity across Scales and Species." *Neuron*. doi.org/10.1016/j.neuron.2025.02.009.

Di Paolo, L. D., White, B., Guénin-Carlut, A., Constant, A., Clark, A. 2024. "Active Inference Goes to School: The Importance of Active Learning in the Age of Large Language Models." *Philosophical Transactions* 379(1911): 20230148.

Doya, K. (ed.). 2007. *Bayesian Brain: Probabilistic Approaches to Neural Coding*. Cambridge, MA: MIT Press.

Dretske, F. 1988. *Explaining Behavior*. Cambridge, MA: MIT Press.

Dupuy, J.-P. 2009. *On the Origins of Cognitive Science*. Cambridge, MA: MIT Press.

Durrett, R. 2019. *Probability: Theory and Examples*. Cambridge: Cambridge University Press.

Edelman, G. M. 1987. *Neural Darwinism: The Theory of Neuronal Group Selection*. New York: Basic Books.

van Es, T. 2020. "Living Models or Life Modelled? On the Use of Models in the Free-Energy Principle." *Adaptive Behavior* 9(3): 315–329.

Evans, M., Rosenthal, J. 2009. *Probability and Statistics: The Science of Uncertainty*. New York: Freeman.

Fabry, R. 2020. "Into the Dark Room: A Predictive Processing Account of Major Depressive Disorder." *Phenomenology and the Cognitive Sciences* 19(4): 685–704.

Fasoli, M. 2021. "The Overuse of Digital Technologies: Human Weaknesses, Design Strategies and Ethical Concerns." *Philosophy and Technology* 34: 1409–1427.

Fasoli, M., Piredda, G. 2023. *Filosofia, tecnologia e scienze della mente*. Bologna: Il Mulino.

Feldman Barrett, L. 2017. *How Emotions Are Made*. New York: Harcourt.

Feldman, H., Friston. K. 2010. "Attention, Uncertainty, and Free-Energy." *Frontiers in Human Neuroscience*. https://doi.org/10.3389/fnhum.2010.00215.

Fields, C., Friston, K., Glazebrook, J. F., Levin, M. 2022. "A Free Energy Principle for Generic Quantum Systems." *Progress in Biophysics and Molecular Biology* 173: 36–59.

Floridi, L. 2019. *The Logic of Information*. Oxford: Oxford University Press.

von Foerster, H. 2003. "On Self-Organizing Systems and Their Environments." In *Understanding Understanding: Essays on Cybernetics and Cognition*, ed. H. von Foerster, 1–19. Berlin: Springer.

Fonagy, P., Bateman, A. 2011. "The Widening Scope of Mentalizing: A Discussion." *Psychology and Psychotherapy: Theory, Research and Practice* 84(1): 98–110.

Fong, C., Law, W., Uka, T., Koike, S. 2020. "Auditory Mismatch Negativity under Predictive Coding Framework and Its Role in Psychotic Disorders." *Frontiers in Psychiatry* 11: 557932. https://doi.org/10.3389/fpsyt.2020.557932.

Fox, S. 2021. "Active Inference: Applicability to Different Types of Social Organization Explained through Reference to Industrial Engineering and Quality Management." *Entropy* 23(2): 198–220.

Frankfurt, H. 1978. "The Problem of Action." *American Philosophical Quarterly* 15(2): 157–162.

Friston, K. 2005. "A Theory of Cortical Responses." *Philosophical Transactions of the Royal Society of London B: Biological Sciences* 360(1456): 815–836.

Friston, K. 2008. "Hierarchical Models in the Brain." *PLOS Computational Biology* 4(11): e1000211.

Friston, K. 2009. "The Free-Energy Principle: A Rough Guide to the Brain?" *Trends in Cognitive Sciences* 13(7): 293–301.

Friston, K. 2010a. "The Free-Energy Principle: A Unified Brain Theory?" *Nature Reviews Neuroscience* 11: 127–138.

Friston, K. 2010b. "Is the Free-Energy Principle Neurocentric?" *Nature Reviews Neuroscience* 11: 605.

Friston, K. 2011. "What Is Optimal about Motor Control?" *Neuron* 72(3): 488–498.

Friston, K. 2013. "Life as We Know It." *Journal of the Royal Society Interface* 10(86): 20130475.

Friston, K. 2017. "Precision Psychiatry." *Biological Psychiatry: Cognitive Neuroscience and Neuroimaging* 2(8): 640–643.

Friston, K. 2019. "A Free Energy Principle for a Particular Physics." *arXiv preprint arXiv:1906.10184.*

Friston, K. 2022. "The Ultimate Trick? Comment on: 'The Markov Blanket Trick: On the Scope of the Free Energy Principle and Active Inference' by Raja et al." *Physics of Life Reviews* 43:10–16.

Friston, K. 2024. "Cultural Mechanics. Comment on: 'To Copy or Not to Copy? That Is the Question! From Chimpanzees to the Foundation of Human Technological Culture' by H. Manrique, and M. Walker." *Physics of Life Reviews* 46: 76–79.

Friston, K., Da Costa, L., Hafner, D., Hesp, C., Parr, T. 2020. "Sophisticated Inference." *arXiv preprint arXiv:2006.04120.*

Friston, K., Da Costa, L., Sakthivadivel, D. A., Heins, C., Pavliotis, G. A., Ramstead, M., Parr, T. 2023a. "Path Integrals, Particular Kinds, and Strange Things." *Physics of Life Reviews* 47: 35–62.

Friston, K., Daunizeau J., Kiebel, S. 2009. "Reinforcement Learning or Active Inference?" *PLOS ONE* 4(7): e6421.

Friston, K., FitzGerald, T., Rigoli, F., Schwartenbeck, P., Pezzulo, G. 2017. "Active Inference: A Process Theory." *Neural Computation* 29(1): 1–49.

Friston, K., Friedman, D., Constant, A., Knight, V., Fields, C., Parr, T., Campbell, J. 2023b. "A Variational Synthesis of Evolutionary and Developmental Dynamics." *Entropy* 25: 964.

Friston, K., Frith, C. 2015a. "Active Inference, Communication and Hermeneutics." *Cortex: A Journal Devoted to the Study of the Nervous System and Behavior* 68: 129–143.

Friston, K., Frith, C. 2015b. "A Duet for One." *Consciousness and Cognition* 36: 390–405.

Friston, K., Kilner, J., Harrison, L. 2006. "A Free Energy Principle for the Brain." *Journal of Physiology—Paris* 100(1–3): 70–87.

Friston, K., Mattout, J., Kilner, J. 2011. "Action Understanding and Active Inference." *Biological Cybernetics* 104(1): 137–160.

Friston, K. J., Ramstead, M. J., Kiefer, A., Tschantz, A., Buckley, C. L., Albarracin, M., Pitliya, R. J., Heins, C., Klein, B., Millidge, B., Sakthivadivel, D. A., St Clere Smithe, T., Koudahl, M., Tremblay, S. E., Petersen, C., Fung, K., Fox, J. G., Swanson, S., Mapes, D., René, G. 2024. "Designing Ecosystems of Intelligence from First Principles." *Collective Intelligence* 3(1). https://doi.org/10.1177/26339137231222481.

Friston, K., Schwartenbeck, P., Fitzgerald, T., Moutoussis, M., Behrens, T., Dolan, R. 2013. "The Anatomy of Choice: Active Inference and Agency." *Frontiers in Human Neuroscience* 25(7): 598.

Friston, K., Shiner, T., FitzGerald, T., Galea, J., Adams, R. 2012a. "Dopamine, Affordance and Active Inference." *PLOS Computational Biology* 8(1). https://doi.org/10.1371/journal.pcbi.1002327.

Friston, K., Thornton, C., Clark, A. 2012b. "Free-Energy Minimization and the Dark-Room Problem." *Frontiers in Psychology* 3: 130. https://doi.org/10.3389/fpsyg.2012.00130.

Frith, C. 2007. *Making Up the Mind.* New York: Blackwell.

Fuchs, T. 2005. "Corporealized and Disembodied Minds." *Philosophy, Psychiatry, & Psychology* 12(2): 95–107.

Gell, A. 1998. *Art and Agency.* Oxford: Clarendon Press.

Gelman, A., Carlin, J. B., Stern, H. S., Rubin, D. 1995. *Bayesian Data Analysis.* New York: Chapman and Hall/CRC.

Gershman, S. 2019. "What Does the Free Energy Principle Tell Us about the Brain?" *Neurons, Behavior, Data Analysis, and Theory* 2(3): 1–10.

Gershman, S., Daw, N. 2012. "Perception, Action and Utility: The Tangled Skein." In *Principles of Brain Dynamics*, ed. S. Gershman, N. Daw, 293–312. Cambridge, MA: MIT Press.

Gibson, J. 2014. "The Theory of Affordances" (1979). In *The People, Place, and Space Reader*, ed. J. Gieseking, W. Mangold, C. Katz, S. Low, S. Saegert, 56–60. London: Routledge.

Giger, J. C., Piçarra, N., Pochwatko, G., Almeida, N., Almeida, A. S. 2025. "Intention to Work with Social Robots." *Multimodal Technologies and Interaction* 9(2): 9. https://doi.org/10.3390/mti9020009.

Gilbert, J., Wusinich, C., Zarate, J. 2022. "A Predictive Coding Framework for Understanding Major Depression." *Frontiers in Human Neuroscience* 16: 787495.

Gillath, O., Karantzas, G., Selcuk, E. 2017. "A Net of Friends: Investigating Friendship by Integrating Attachment Theory and Social Network Analysis." *Personality and Social Psychology Bulletin* 43(11): 1546–1565.

Glickman, M., Sharot, T. 2024. "How Human–AI Feedback Loops Alter Human Perceptual, Emotional and Social Judgements." *Nature Human Behaviour* 9: 345–359.

Goldman, A. I. 2006. *Simulating Minds.* Oxford: Oxford University Press.

Gonçalves, M., Cash, P. 2021. "The Life Cycle of Creative Ideas: Towards a Dual-Process Theory of Ideation." *Design Studies* 72: 100988.

Gong, T., Gavriilidis, P., Ji, R., Huang, C., Alexandropoulos, G. C., Wei, L., Yuen, C. 2023. "Holographic MIMO Communications: Theoretical Foundations, Enabling Technologies, and Future Directions." *IEEE Communications Surveys & Tutorials* 26(1): 196–257.

Gray, K., Wegner, D. 2012. "Feeling Robots and Human Zombies: Mind Perception and the Uncanny Valley." *Cognition* 125(1): 125–130.

Haken, H. 1983. *Synergetics: An Introduction*. Berlin: Springer.

Halegoua, G. 2020. *Smart Cities*. Cambridge, MA: MIT Press.

Han, D., Mulyana, B., Stankovic, V., Cheng, S. 2023. "A Survey on Deep Reinforcement Learning Algorithms for Robotic Manipulation." *Sensors* 23(7): 3762.

Harman, G. 2016. *Immaterialism*. Cambridge: Polity Press.

Harman, G. 2018. *Object-Oriented Ontology*. London: Penguin Random House.

Harris, D., Arthur, T., Broadbent, D., Wilson, M., Vine, S., Runswick, O. 2022. "An Active Inference Account of Skilled Anticipation in Sport." *Sports Medicine* 52(9): 2023–2038.

Haugeland, J. 1985. *Artificial Intelligence: The Very Idea*. Cambridge, MA: MIT Press.

Heersmink, R. 2017. "Distributed Cognition and Distributed Morality: Agency, Artifacts and Systems." *Science and Engineering Ethics* 23(2): 431–448.

Heersmink, R. 2021. "Varieties of Artifacts: Embodied, Perceptual, Cognitive, and Affective." *Topics in Cognitive Science* 13(4): 573–596.

Heidegger, M. 1927. *Sein und Zeit*. Tübingen: Max Niemeyer Verlag.

Heins, C., Millidge, B., Da Costa, L., Mann, R. P., Friston, K., Couzin, I. 2024. "Collective Behavior from Surprise Minimization." *Proceedings of the National Academy of Sciences* 121(17): e2320239121.

Hesp, C., Smith, R., Parr, T., Allen, M., Friston, K., Ramstead, M. 2021. "Deeply Felt Affect: The Emergence of Valence in Deep Active Inference." *Neural Computation* 33(2): 398–446.

Hinton, G. E., van Camp, D. 1993. "Keeping Neural Networks Simple." In *ICANN '93: Proceedings of the International Conference on Artificial Neural Networks (Amsterdam, the Netherlands 13–16 September 1993)*, ed. S. Gielen, B. Kappen, 134–147. Berlin: Springer.

Hobbis, S. K., Hobbis, G. 2024. "A Sociotechnical Approach to Smartphone Research." *Media International Australia*. https://doi.org/10.1177/1329878X241253011.

Hohwy, J. 2013. *The Predictive Mind*. Oxford: Oxford University Press.

Hohwy, J. 2016. "The Self-Evidencing Brain." *Noûs* (50): 259–285.

Hohwy, J. 2020. "New Directions in Predictive Processing." *Mind & Language* 35(2): 209–223.

Holmes, J. 2014. *John Bowlby and Attachment Theory*. London: Routledge.

Hornborg, A. 2017. "Artifacts Have Consequences, Not Agency." *European Journal of Social Theory* 20(1): 95–110.

Houkes, W., Vermaas, P. 2010. *Technical Functions: On the Use and Design of Artifacts*. Berlin: Springer.

Hutchins, E. 1995. *Cognition in the Wild*. Cambridge, MA: MIT Press.

Hutchinson-Wong, N., Glue, P., Adhia, D., de Ridder, D. 2025. "How Does Depressive Cognition Develop?" *Psychological Review* 132(2): 442–469.

Ihde, D. 1986. *Experimental Phenomenology*. Albany: State University of New York Press.

Ihde, D. 1990. *Technology and the Lifeworld*. Bloomington: Indiana University Press.

Ihde, D. 2023. *The Critical Ihde*. Albany: State University of New York Press.

Ingold, T. 2013. *Making: Anthropology, Archaeology, Art and Architecture*. London: Routledge.

Isomura, T., Kotani, K., Jimbo, Y., Friston, K. 2023. "Experimental Validation of the Free-Energy Principle with in Vitro Neural Networks." *Nature Communications* 14(1): 4547.

Jentzen, A., Kuckuck, B., von Wurstemberger, P. 2023. "Mathematical Introduction to Deep Learning: Methods, Implementations, and Theory." *arXiv preprint arXiv:2310.20360*.

Joyce, J. 2011. "Kullback–Leibler Divergence." In *International Encyclopedia of Statistical Science*, ed. M. Lovric, 720–722. Berlin: Springer.

Kabadayi, F. 2024. "Smartphone Addiction, Depression, Distress, Eustress, Loneliness, and Sleep Deprivation in Adolescents: A Latent Profile and Network Analysis Approach." *BMC Psychology* 12(1): 608.

Kahneman, D. 2011. *Thinking, Fast and Slow*. New York: Macmillan.

Kallinikos, J. 2002. "Reopening the Black Box of Technology Artifacts and Human Agency." *LSE Research Online*. eprints.lse.ac.uk/190/1/Kallinikos-ICIS.pdf.

Kaufmann, R., Gupta, P., Taylor, J. 2021. "An Active Inference Model of Collective Intelligence." *Entropy* 23(7): 830.

Keller, E. 2008. "Organisms, Machines, and Thunderstorms: A History of Self-Organization, Part One." *Historical Studies in the Natural Sciences* 38(1): 45–75.

Kirchhoff, M., Parr, T., Palacios, E., Friston, K., Kiverstein, J. 2018. "The Markov Blankets of Life: Autonomy, Active Inference and the Free-Energy Principle." *Journal of the Royal Society Interface*. doi.org/10.1098/rsif.2017.0792.

Klein, C. 2018. "What Do Predictive Coders Want?" *Synthese* 195(6): 2541–2557.

Knill, D., Pouget, A. 2004. "The Bayesian Brain: The Role of Uncertainty in Neural Coding and Computation." *Trends in Neurosciences* 27(12): 712–719.

Koole, S., Tschacher, W. 2016. "Synchrony in Psychotherapy: A Review and an Integrative Framework for the Therapeutic Alliance." *Frontiers in Psychology* (7): 862–875.

Kroes, P., Meijers, A. (eds.). 2000. *The Empirical Turn in the Philosophy of Technology*. New York: JAI/Elsevier.

Kube, T., Berg, M., Kleim, B., Herzog, P. 2020. "Rethinking Post-Traumatic Stress Disorder: A Predictive Processing Perspective." *Neuroscience & Biobehavioral Reviews* 113: 448–460.

Kukkonen, K. 2020. *Probability Designs*. Oxford: Oxford University Press.

Kukkonen, K. 2024. "Designs on Consciousness: Literature and Predictive Processing." *Philosophical Transactions of the Royal Society* 379: 20220423. https://doi.org/10.1098/rstb.2022.0423.

Latour, B. 1988. *The Pasteurization of France*. Cambridge, MA: Harvard University Press.

Latour, B. 1992. "Where Are the Missing Masses? The Sociology of a Few Mundane Artifacts." In *Shaping Technology/Building Society: Studies in Sociotechnical Change*, ed. W. Bijker, J. Law, 134–156. Cambridge, MA: MIT Press.

Latour, B. 1993a. "Portrait de Gaston Lagaffe en philosophe des techniques." In *La Clef de Berlin et autres leçons d'un amateur de sciences*, ed. B. Latour, 23–38. Paris: La Découverte.

Latour, B. 1993b. "La Clef de Berlin." In *Petites Leçons de sociologie des sciences*, ed. B. Latour, 45–56. Paris: La Découverte.

Latour, B. 1993c. "Le Groom est en grève: pour l'amour de Dieu, fermez la porte." In *Petites Leçons de sociologie des sciences*, ed. B. Latour, 56–67. Paris: La Découverte.

Latour, B. 1993d. *We Have Never Been Modern*. Cambridge, MA: Harvard University Press.

Latour, B. 1994. "On Technical Mediation." *Common Knowledge* 3(2): 29–64.

Latour, B. 1999. *Pandora's Hope: Essay on the Reality of Science Studies*. Cambridge, MA: Harvard University Press.

Latour, B. 2007. *Reassembling the Social*. Oxford: Oxford University Press.

Laukkonen, R., Friston, K., Chandaria, S. 2025. "A Beautiful Loop: An Active Inference Theory of Consciousness." *Neuroscience and Biobehavioral Reviews* 176 https://doi.org/10.1016/j.neubiorev.2025.106296.

Law, J. 1987. "Technology Transfer and Social Engineering." *Social Studies of Science* 17(3): 564–569.

Law, J. 2002. *Aircraft Stories: Decentering the Object in Technoscience*. Durham, NC: Duke University Press.

Law, J. 2008. "Actor Network Theory and Material Semiotics." In *The New Blackwell Companion to Social Theory*, ed. B. Turner, 141–158. London: Blackwell.

Lee, J. H., Ostwald, M. 2022. "The Relationship between Divergent Thinking and Ideation in the Conceptual Design Process." *Design Studies* 79: 101089.

Levin, M. 2022. "Technological Approach to Mind Everywhere: An Experimentally-Grounded Framework for Understanding Diverse Bodies and Minds." *Frontiers in Systems Neuroscience* 16: 768201.

Levine, S. 2018. "Reinforcement Learning and Control as Probabilistic Inference: Tutorial and Review." *arXiv:1805.00909.*

Liberati, N. 2016. "Augmented Reality and Ubiquitous Computing: The Hidden Potentialities of Augmented Reality." *AI & Society* 31: 17–28. https://doi.org/10.1007/s00146-014-0543-x.

Lindgren, S. 2024. *Critical Theory of AI.* London: Polity.

Lundy, B. 2003. "Father– and Mother–Infant Face-to-Face Interactions: Differences in Mind-Related Comments and Infant Attachment?" *Infant Behavior and Development* 26(2): 200–212.

Luyten, P., Campbell, C., Allison, E., Fonagy, P. 2020. "The Mentalizing Approach to Psychopathology: State of the Art and Future Directions." *Annual Review of Clinical Psychology* 16(1): 297–325.

MacDorman, K. 2024. "Does Mind Perception Explain the Uncanny Valley? A Meta-Regression Analysis and (De)Humanization Experiment." *Computers in Human Behavior: Artificial Humans* 2(1): 100065.

MacDorman, K. 2025. "Chikamatsu, Mori, and the Uncanny Valley." *i-Perception* 16(1): 20416695251317469.

MacKay, D. J. 2003. *Information Theory, Inference and Learning Algorithms.* Cambridge: Cambridge University Press.

Maisto, D., Gregoretti, F., Friston, K., Pezzulo, G. 2025. "Active Inference Tree Search in Large POMDPs." *Neurocomputing* 623: 129319.

Malafouris, L. 2013. *How Things Shape the Mind.* Cambridge, MA: MIT Press.

Mann, S., Pain, R., Kirchhoff, M. 2022. "Free Energy: A User's Guide." *Biology & Philosophy* 37(4): 33–56.

Manrique, H., Friston, K., Walker, M. 2024. "'Snakes and Ladders' in Paleoanthropology: From Cognitive Surprise to Skillfulness a Million Years Ago." *Physics of Life Reviews* 49: 40–70.

Manrique, H., Walker, M. 2017. *Early Evolution of Human Memory. Great Apes, Toolmaking, and Cognition.* London: Palgrave.

Manrique, H., Walker, M. 2023. "To Copy or Not to Copy? That Is the Question! From Chimpanzees to the Foundation of Human Technological Culture." *Physics of Life Reviews* 45: 6–24.

Maroto-Gómez, M., Alonso-Martín, F., Malfaz, M., Castro-González, Á., Castillo, J., Salichs, M. 2023. "A Systematic Literature Review of Decision-Making and Control Systems for Autonomous and Social Robots." *International Journal of Social Robotics* 15(5): 745–789.

Marr, D. 1982. *Vision: A Computational Investigation into the Human Representation and Processing of Visual Information*. San Francisco, CA: Freeman.

Mattingly, J., Cibralic, B. 2025. *Machine Agency*. Cambridge, MA: MIT Press.

McCulloch, W. 2016. *Embodiments of Mind*. Cambridge, MA: MIT Press.

McElreath, R. 2020. *Statistical Rethinking*. New York: Chapman and Hall/CRC.

McNeill, W. 1976. *Plagues and Peoples*. New York: Anchor Press.

Meins, E., Fernyhough, C., Wainwright, R., Das Gupta, M., Fradley, E., Tuckey, M. 2002. "Maternal Mind–Mindedness and Attachment Security as Predictors of Theory of Mind Understanding." *Child Development* 73(6): 1715–1726.

Menary, R. (ed.). 2010. *The Extended Mind*. Cambridge, MA: MIT Press.

Mendonça, D., Curado, M., Gouveia, S. (eds.). 2020. *The Philosophy and Science of Predictive Processing*. London: Bloomsbury.

Metzinger, T., Wiese, W. (eds.). 2017. *Philosophy and Predictive Processing*. Frankfurt am Main: MIND Group.

Millidge, B., Seth, A., Buckley, C. 2021. "A Mathematical Walkthrough and Discussion of the Free Energy Principle." *arXiv, abs/2108.13343*.

Mitcham, C. 2013. "Agency in Humans and in Artifacts: A Contested Discourse." In *The Moral Status of Technical Artefacts*, ed. P. Kroes, P.-P. Verbeek, 11–29. Dordrecht: Springer.

Modirshanechi, A., Brea, J., Gerstner, W. 2022. "A Taxonomy of Surprise Definitions." *Journal of Mathematical Psychology* 110: 102712.

Mori, M. 1970. "The Uncanny Valley." *Energy* 7(4): 33–35.

Mossner, C., Walter, S. 2025. "Scaffolded Affective Harm: What Is It and (How) Can We Do Something about It?" *Topoi*. https://doi.org/10.1007/s11245-024-10136-6.

Murphy, K. 2012. *Machine Learning: A Probabilistic Perspective*. Cambridge, MA: MIT Press.

Murray, S., Kersten, D., Olshausen, B., Schrater, P., Woods, D. 2002. "Shape Perception Reduces Activity in Human Primary Visual Cortex." *PNAS* 99(23): 15164–15169.

Nave, K. 2025. *A Drive to Survive: The Free Energy Principle and the Meaning of Life*. Cambridge, MA: MIT Press.

Nave, K., Deane, G., Miller, M., Clark, A. 2020. "Wilding the Predictive Brain." *WIREs Cognitive Sciences* e1542. https://doi.org/10.1002/wcs.1542.

Nehrer, S., Ehrenreich Laursen, J., Heins, C., Friston, K., Mathys, C., Thestrup Waade, P. 2025. "Introducing ActiveInference.jl: A Julia Library for Simulation and Parameter Estimation with Active Inference Models." *Entropy* 27(1): 62. https://doi.org/10.3390/e27010062.

Nelson, G., E. Stolterman. 2012. *The Design Way: Intentional Change in an Unpredictable World*. Cambridge, MA: MIT Press.

Nicolis, G., Prigogine, I. 1977. *Self-Organization in Nonequilibrium Systems*. New York: Wiley-Interscience.

Norman, D. 2004. *Emotional Design*. New York: Basic Books.

Norman, D. 2013. *The Design of Everyday Things*. Cambridge, MA: MIT Press.

Norman, D. 2016. *Living with Complexity*. Cambridge, MA: MIT Press.

Novica, D. R., Wianto, E., Andrade Campos, S. 2023. "Drawing and Ideation Process in Design Education: A Systematic Literature Review." *Cogent Arts & Humanities* 10(1): 2219487.

Oh, W., Heo, Y. 2024. "Exploring the Link between Smartphone Overdependence, Depression, and Suicidal Behaviors through the Mediating Effect of Lifestyle Risk Behaviors among South Korean Adolescents: A Cross-Sectional Study Using National Big Data." *Journal of Pediatric Health Care* 38(3): 298–309.

Ororbia, A., Friston, K. 2024. "Mortal Computation: A Foundation for Biomimetic Intelligence." *arXiv:2311.09589*.

Panksepp, J. 1998. *Affective Neuroscience*. Oxford: Oxford University Press.

Panksepp, J., Biven, L. 2012. *The Archaeology of Mind*. New York: Norton.

Parr, T. 2021. "Message Passing and Metabolism." *Entropy* 23: 606. https://doi.org/10.3390/e23050606.

Parr, T., Benrimoh, D., Vincent, P., Friston, K. 2018. "Precision and False Perceptual Inference." *Frontiers in Integrative Neuroscience* 12: 39. https://doi.org/10.3389/fnint.2018.00039.

Parr, T., Da Costa, L., Friston, K. 2019. "Markov Blankets, Information Geometry, and Stochastic Thermodynamics." *Philosophical Transactions* 378: 2019059.

Parr, T., Friston, K. 2017. "Uncertainty, Epistemics and Active Inference." *Journal of the Royal Society Interface* 14: 20170376. http://dx.doi.org/10.1098/rsif.2017.0376.

Parr, T., Friston, K. 2018. "The Anatomy of Inference: Generative Models and Brain Structure." *Frontiers in Computational Neuroscience* 13: 90. https://doi.org/10.3389/fncom.2018.00090.

Parr, T., Pezzulo, G., Friston, K. 2022. *Active Inference: The Free Energy Principle in Mind, Brain, and Behavior*. Cambridge, MA: MIT Press.

Parsons, G. 2015. *The Philosophy of Design*. Cambridge: Polity Press.

Pearl, J. 1988. *Probabilistic Reasoning in Intelligent Systems: Networks of Plausible Inference*. New York: Kaufmann.

Petitot, J. 1992. *Physique du sens: de la théorie des singularités aux structures sémio-narratives*. Paris: CNRS.

Pezzulo, G. 2013. "Why Do You Fear the Bogeyman? An Embodied Predictive Coding Model of Perceptual Inference." *Cognitive, Affective, and Behavioral Neuroscience* 14(3): 902–911.

Pezzulo, G., Cisek, P. 2016. "Navigating the Affordance Landscape: Feedback Control as a Process Model of Behavior and Cognition." *Trends in Cognitive Sciences* 20: 414–424.

Pezzulo, G., Parr, T., Cisek, P., Clark, A., Friston, K. 2024a. "Generating Meaning: Active Inference and the Scope and Limits of Passive AI." *Trends in Cognitive Sciences* 28(2): 97–112.

Pezzulo, G., Parr, T., Friston, K. 2021. "The Evolution of Brain Architectures for Predictive Coding and Active Inference." *Philosophical Transactions of the Royal Society B* 377: 20200531.

Pezzulo, G., Parr, T., Friston, K. 2024b. "Active Inference as a Theory of Sentient Behavior." *Biological Psychology* 186: 108741.

Pezzulo, G., Rigoli, F., Friston, K. 2015. "Active Inference, Homeostatic Regulation and Adaptive Behavioural Control." *Progress in Neurobiology* 134: 17–35.

Pfaff, D. 2005. *Brain Arousal and Information Theory*. Cambridge, MA: Harvard University Press

Pikovsky, A., Rosenblum, M., Kurths, J. 2001. *Synchronization: A Universal Concept in Nonlinear Sciences*. Cambridge: Cambridge University Press.

Pio-Lopez, L., Kuchling, F., Tung, A., Pezzulo, G., Levin, M. 2022. "Active Inference, Morphogenesis, and Computational Psychiatry." *Frontiers in Computational Neuroscience* 16: 988977. https://doi.org/10.3389/fncom.2022.988977.

Piredda, G. 2020. "What Is an Affective Artifact? A Further Development in Situated Affectivity." *Phenomenology and the Cognitive Sciences* 19: 549–567.

Possati, L. M., Aydin, C. 2025. "To Repair or Not to Repair? Disclosing the Subject in Technical Mediation." *Philosophy & Technology* 38(3). https://doi.org/10.1007/s13347-025-00930-4.

Prigogine, I. 1996. *La Fin des certitudes*. Paris: Odile Jacob.

Prunotto, A. 2022. "Towards an Integrated Theory of Mediation." *Explorations in Media Ecology* 21: 231–249. https://doi.org/10.1386/eme_00131_1.

Ramstead, M., Badcock, P., Friston, K. 2018. "Answering Schrödinger's Question: A Free Energy Formulation." *Physics of Life Reviews* (24): 1–16.

Ramstead, M., Constant, A., Badcock, P., Friston, K. 2019a. "Variational Ecology and the Physics of Sentient Systems." *Physics of Life Reviews* 31: 188–205.

Ramstead, M., Hesp, C., Tschantz, A., Smith, R., Constant, A., & Friston, K. 2021. "Neural and Phenotypic Representation under the Free Energy Principle." *Neuroscience & Biobehavioral Reviews* 120: 109–122.

Ramstead, M., Kirchhoff, M., Constant, A., Friston, K. 2019b. "Multiscale Integration: Beyond Internalism and Externalism." *Synthese* 198: 41–70. https://doi.org/10.1007/s11229-019-02115-x.

Ramstead, M., Kirchhoff, M., Friston, K. 2019c. "A Tale of Two Densities: Active Inference Is Enactive Inference." *Adaptive Behavior* 28(4): 181–204.

Ramstead, M., Sakthivadivel, D., Friston, K. 2024. "An Approach to Non-Equilibrium Statistical Physics Using Variational Bayesian Inference." *arXiv preprint arXiv:2406.11630.*

Ramstead, M., Sakthivadivel, D., Heins, C., Koudahl, M., Millidge, B., Da Costa, L., Klein, B., Friston, K. 2023. "On Bayesian Mechanics: A Physics of and by Beliefs." *Interface Focus* 13: 20220029. https://doi.org/10.1098/rsfs.2022.0029.

Ramstead, M., Veissière, S., Kirmayer, L. 2016. "Cultural Affordances: Scaffolding Local Worlds through Shared Intentionality and Regimes of Attention." *Frontiers in Psychology* 7: 1090. https://doi.org/10.3389/fpsyg.2016.01090.

Ratcliffe, M. 2015. *Experiences of Depression: A Study in Phenomenology*. Oxford: Oxford University Press.

Reddy, V. 2008. *How Infants Know Minds*. Cambridge, MA: Harvard University Press.

Rendell, L., Fogarty, L., Hoppitt, W., Morgan, T., Webster, W., Laland, K. 2011. "Cognitive Culture: Theoretical and Empirical Insights into Social Learning Strategies." *Trends in Cognitive Sciences* 15(2): 68–76.

Rezaei Khavas, Z., Kotturu, M., Ahmadzadeh, S., Robinette, P. 2024. "Do Humans Trust Robots That Violate Moral Trust?" *ACM Transactions on Human–Robot Interaction* 13(2): 1–30.

Riva, G., Mancini, T. 2023. *Psicologia dei media digitali*. Bologna: Il Mulino.

Robinson, J., Corcoran, A., Whyte, C., Sárközy, A., Seth, A., Kovács, G. 2025. "The Role of Active Inference in Conscious Awareness." *PLoS One* 20(12): e0328836. https://doi.org/10.1371/journal.pone.0328836.

Roesler, E., Vollmann, M., Manzey, D., Onnasch, L. 2024. "The Dynamics of Human–Robot Trust Attitude and Behavior: Exploring the Effects of Anthropomorphism and Type of Failure." *Computers in Human Behavior* 150: 108008.

Rosenberg, R., Verbeek, P.-P. (eds.). 2015. *Postphenomenological Investigations*. Lanham, MD: Lexington Books.

Rosenfeld, E., Anderson, J. (eds.). 1989. *Neurocomputing. Foundations of Research.* Cambridge, MA: MIT Press.

Sakthivadivel, D. 2022a. "Weak Markov Blankets in High-Dimensional, Sparsely-Coupled Random Dynamical Systems." *arXiv:2207.07620.*

Sakthivadivel, D. 2022b. "Towards a Geometry and Analysis for Bayesian Mechanics." *arXiv preprint arXiv:2204.11900.*

Sakthivadivel, D. 2022c. "A Worked Example of the Bayesian Mechanics of Classical Objects." In *International Workshop on Active Inference*, ed. C. Buckley, D. Cialfi, P. Lanillos, M. Ramstead, N. Sajid, H. Shimazaki, T. Verbelen, 298–318. Cham: Springer.

Sandved-Smith, L., Hesp, C., Mattout, J., Friston, K., Lutz, A., Ramstead, M. 2021. "Towards a Computational Phenomenology of Mental Action: Modelling Meta-Awareness and Attentional Control with Deep Parametric Active Inference." *Neuroscience of Consciousness*. https://doi.org/10.1093/nc/niab018.

Schrodinger, E. 1944. *What Is life? The Physical Aspect of the Living Cell.* Cambridge: Cambridge University Press.

Schulkin, J., Sterling, P. 2019. "Allostasis: A Brain-Centered, Predictive Mode of Physiological Regulation." *Trends in Neurosciences* 42(10): 740–752.

Sclove, R. 1995. *Democracy and Technology*. New York: Guilford Press.

Seth, A. 2015. "The Cybernetic Bayesian Brain." In *Open Mind*, ed. T. Metzinger, J. Windt, 1–24. Frankfurt am Main: Open Mind.

Seth, A. 2021. *Being You: A New Science of Consciousness*. London: Faber & Faber.

Shannon, C. E. 1948. "A Mathematical Theory of Communication." *Bell System Technical Journal* 27(3): 379–423.

Sharma, M., Rao, G., Benegal, V., Thennarasu, K., Thomas, D. 2017. "Technology Addiction Survey: An Emerging Concern for Raising Awareness and Promotion of Healthy Use of Technology." *Indian Journal of Psychological Medicine* 39(4): 495–499.

Shavit, Y., Agarwal, S., Brundage, M. 2023. "Practices for Governing Agentic AI Systems." *OpenAI.* https://openai.com/index/practices-for-governing-agentic-ai-systems/.

Sherer, J., Levounis, P. 2022. "Technological Addictions." *Current Psychiatry Reports* 24(9): 399–406.

Shvetsov, A. V., Alsamhi, S. H. 2024. "When Holographic Communication Meets Metaverse: Applications, Challenges and Future Trends." *IEEE Access*. 10.1109/ACCESS.2024.3514576.

Simon, H. 1981. *The Sciences of the Artificial.* Cambridge, MA: MIT Press.

Simondon, G. 2005. *L'Individuation à la lumière des notions de forme et d'information.* Grenoble: Éditions Jérôme Millon.

Sims, A. 2016. "A Problem of Scope for the Free Energy Principle as a Theory of Cognition." *Philosophical Psychology* 29(7): 967–980.

Sloterdijk, P. 2011–2016. *Spheres 1–3*. Cambridge, MA: MIT Press.

Smith, R., Friston, K., Whyte, C. 2022. "A Step-by-Step Tutorial on Active Inference and Its Application to Empirical Data." *Journal of Mathematical Psychology* 107: 102632. https://doi.org/10.1016/j.jmp.2021.102632.

Smith, R., Kuplicki, R., Feinstein, J., Forthman, K., Stewart, J., Paulus, M., et al. 2020a. "A Bayesian Computational Model Reveals a Failure to Adapt Interoceptive Precision Estimates across Depression, Anxiety, Eating, and Substance Use Disorders." *PLoS Computational Biology* 14; 16(12): e1008484. https://doi.org/10.1371/journal.pcbi.1008484.

Smith, R., Schwartenbeck, P., Parr, T., Friston, K. 2020b. "An Active Inference Approach to Modeling Structure Learning: Concept Learning as an Example Case." *Frontiers in Computational Neuroscience* 14: 41. https://doi.org/10.3389/fncom.2020.00041.

Solms, M. 2021. *The Hidden Spring: A Journey to the Sources of Consciousness*. London: Profile Books.

Solms, M., Friston, K. 2018. "How and Why Consciousness Arises: Some Considerations from Physics and Physiology." *Journal of Consciousness Studies* 25(5–6): 202–238.

Song, Y., Luximon, Y. 2020. "Trust in AI Agent: A Systematic Review of Facial Anthropomorphic Trustworthiness for Social Robot Design." *Sensors* 20(18): 5087–5097.

Stein, J., MacDorman, K. 2024. "After Confronting One Uncanny Valley, Another Awaits." *Nature Reviews Electrical Engineering* 1(5): 276–277.

Summerfield, C., Trittschuh, E., Monti, J., Mesulam, M., Egner, T. 2008. "Neural Repetition Suppression Reflects Fulfilled Perceptual Expectations." *Nature Neuroscience* 11(9): 1004–1006.

Tenenbaum, J., Kemp, C., Griffiths, T., Goodman, N. 2011. "How to Grow a Mind: Statistics, Structure, and Abstraction." *Science* 331: 1279–1285.

Thestrup Waade, P., Lundbak Olesen, C., Ehrenreich Laursen, J., Nehrer, S., Heins, C., Friston, K., Mathys, C. 2025. "As One and Many: Relating Individual and Emergent Group-Level Generative Models in Active Inference." *Entropy* 27(2): 143. https://doi.org/10.3390/e27020143.

Tomasello, M. 1993. "The Interpersonal Origins of Self Concept." In *The Perceived Self: Ecological and Interpersonal Sources of Self-Knowledge*, ed. U. Neisser, 174–184. Cambridge: Cambridge University Press.

Tomasello, M. 1999. *The Cultural Origins of Human Cognition*. Cambridge, MA: Harvard University Press.

Tomasello, M. 2021. *Becoming Human*. Cambridge, MA: Harvard University Press.

Trevarthen, C. 1979. "Communication and Cooperation in Early Infancy: A Description of Primary Intersubjectivity." In *Before Speech: The Beginning of Human Communication*, ed. M. Bullowa, 321–347. Cambridge: Cambridge University Press.

Tschacher, W., Giersch, A., Friston, K. 2017. "Embodiment and Schizophrenia: A Review of Implications and Applications." *Schizophrenia Bulletin* 43(4): 745–753.

Tschantz, A., Millidge, B., Seth, A., Buckley, C. 2020. "Reinforcement Learning through Active Inference." *arXiv:2002.12636*.

Turing, A. M. 1936. "On Computable Numbers, with an Application to the *Entscheidungsproblem*." *Proceedings of the London Mathematical Society* 2(42): 230–265.

Van de Cruys, S., Van Dessel, P. 2021. "Mental Distress through the Prism of Predictive Processing Theory." *Current Opinion in Psychology* 41: 107–112.

Vasil, J., Badcock, P., Constant, A., Friston, K., Ramstead, M. 2020. "A World unto Itself: Human Communication as Active Inference." *Frontiers in Psychology* 11: 417. https://doi.org/10.3389/fpsyg.2020.00417.

Veissière, S. 2016. "Varieties of Tulpa Experiences." In *Hypnosis and Meditation: Towards an Integrative Science of Conscious Planes*, ed. A. Raz, M. Lifshitz, 55–74. Oxford: Oxford University Press.

Veissière, S., Constant, A., Ramstead, M., Friston, K., Kirmayer, L. 2020. "Thinking through Other Minds: A Variational Approach to Cognition and Culture." *Behavioral and Brain Sciences* 43:e90.

Veissière, S. P., Stendel, M. 2018. "Hypernatural Monitoring: A Social Rehearsal Account of Smartphone Addiction." *Frontiers in Psychology* 9: 141. https://doi.org/10.3389/fpsyg.2018.00141.

Verbeek, P.-P. 2005. *What Things Do: Philosophical Reflections on Technology, Agency, and Design*. University Park: Penn State University Press.

Verbeek, P.-P. 2008. "Obstetric Ultrasound and the Technological Mediation of Morality: A Postphenomenological Analysis." *Human Studies* 31(1): 11–26.

Vial, S. 2017. *L'Être et l'écran*. Paris: Puf.

Walker, M. 2025. "*Homo Informatio*." *Physics of Life Reviews* 55: 98–119. https://doi.org/10.1016/j.plrev.2025.09.007.

Wang, S., Sanches de Oliveira, G., Djebbara, Z., Gramann, K. 2022. "The Embodiment of Architectural Experience: A Methodological Perspective on Neuro-Architecture." *Frontiers in Human Neuroscience* 16: 833528. https://doi.org/10.3389/fnhum.2022.833528.

Wei, R. 2024. "Value of Information and Reward Specification in Active Inference and POMDPs." *arXiv:2408.06542*.

Weisberg, M. 2013. *Simulation and Similarity: Using Models to Understand the World.* Oxford: Oxford University Press.

Wellman, H. 1990. *The Child's Theory of Mind.* Cambridge, MA: MIT Press.

Wellner, G. (ed.). 2020. *Postphenomenology and Media: Essays on Human–Media–World Relations.* London: Lexington Books.

White, B., Miller, M. 2021. "Filtered States: Active Inference, Social Media and Mental Health." In *Machine Learning and Principles and Practice of Knowledge Discovery in Databases*, ed. M. Kamp, 772–783. Cham: Springer. https://doi.org/10.1007/978-3-030-93736-2_54.

Whittaker, J. 2009. *Graphical Models in Applied Multivariate Statistics.* New York: Wiley.

Whittle, A., Spicer, A. 2008. "Is Actor Network Theory Critique?" *Organization Studies* 29(4): 611–629.

Whyte, C., Corcoran, A., Robinson, J., Smith, R., Moran, R., Parr, T., et al. 2024. "On the Minimal Theory of Consciousness Implicit in Active Inference." *arXiv:2410.06633.*

Wiese, W. 2017. *Experienced Wholeness.* Cambridge, MA: MIT Press.

Wiese, W. 2024. "Artificial Consciousness: A Perspective from the Free Energy Principle." *Philosophical Studies* 181: 1947–1970. https://doi.org/10.1007/s11098-024-02182-y.

Winner, L. 1980. "Do Artifacts Have Politics?" *Daedalus* 109: 121–136.

Winnicott, D. 1971. *Playing and Reality.* London: Tavistock.

Woodward, J. 2003. *Making Things Happen.* New York: Oxford University Press.

Xu, Y., Liu, J., Wu, X., Guo, T., Peng, H. 2024. "Reconfigurable Holographic Surface-Assisted Wireless Secrecy Communication System." *Electronics* 13(7): 1359.

Yoshida, W., Dolan, R., Friston, K. 2008. "Game Theory of Mind." *PLOS Computational Biology* 4(12): e1000254. https://doi.org/10.1371/journal.pcbi.1000254.

Yunus, M., Hanapi, N., Kamarudin, Z., Mohd Sufian, S., Ali, A., Ripin, F., Sofian, H. 2025. "Exploring the Landscape of 6G Wireless Communication Technology: A Review." *International Journal of Advanced Computer Science & Applications* 16(5): 123–132.

Zmarz, P., Keller, G. 2016. "Mismatch Receptive Fields in Mouse Visual Cortex." *Neuron* 92(4): 766–772.

INDEX

Publisher contact:
The MIT Press
Massachusetts Institute of Technology
77 Massachusetts Avenue, Cambridge, MA 02139
mitpress.mit.edu

EU Authorised Representative:
Easy Access System Europe, Mustamäe tee 50,
10621 Tallinn, Estonia
gpsr.requests@easproject.com

Printed by Integrated Books International,
United States of America